Springer Monographs in Mathematics

This series publishes advanced monographs giving well-written presentations of the "state-of-the-art" in fields of mathematical research that have acquired the maturity needed for such a treatment. They are sufficiently self-contained to be accessible to more than just the intimate specialists of the subject, and sufficiently comprehensive to remain valuable references for many years. Besides the current state of knowledge in its field, an SMM volume should ideally describe its relevance to and interaction with neighbouring fields of mathematics, and give pointers to future directions of research.

Johanna N. Y. Franklin · Isaac Goldbring ·
Timothy H. McNicholl

Effective Metric Structure Theory

Springer

Johanna N. Y. Franklin
Department of Mathematics
Hofstra University
Hempstead, NY, USA

Isaac Goldbring
Department of Mathematics
University of California
Irvine, CA, USA

Timothy H. McNicholl
Department of Mathematics
Iowa State University
Ames, IA, USA

ISSN 1439-7382 ISSN 2196-9922 (electronic)
Springer Monographs in Mathematics
ISBN 978-3-032-13321-2 ISBN 978-3-032-13322-9 (eBook)
https://doi.org/10.1007/978-3-032-13322-9

This Springer imprint is published by the registered company Springer Nature Switzerland AG
The registered company address is: Gewerbestrasse 11, 6330 Cham, Switzerland

To my family, both birth and chosen.

—Johanna N. Y. Franklin

To Karina, Kaylee, and Daniella, for their neverending love and support.

—Isaac Goldbring

Firstly to my wife Susan who unfailingly stood by my side through the many struggles that led to this point in my career. Secondly, to all the mathematicians who advised, taught, and encouraged me along the way: my advisor Valentina Harizanov, Julia Knight, Alec Matheson, Valentin Andreev, Dale Daniel, Jack Lutz, Carl Jockusch, Steffen Lempp, and the late Paul Schupp.

—Timothy H. McNicholl

Preface

We, the authors of this book, are logicians. The first and third authors (Franklin and McNicholl) are computability theorists, and the second author (Goldbring) is a model theorist. We are united by an intense interest in interactions between mathematical logic and analysis, though we approach this topic from different perspectives. All of us, therefore, have been drawn to the field of effective metric structure theory, which combines elements of model theory, computability theory, and various branches of analysis (though primarily functional analysis).

The seeds of effective metric structure theory arguably lie in the work of Pour-El and Richards, though certainly much of the development in the field has occurred in the past ten years. When the third author began working in this area, there were only a few results obtained by a fairly small company of scholars. Since then, interest has increased greatly along with the number and breadth of discoveries. In particular, the field has drawn some interest from model theorists as well as from an increasing number of graduate students in various areas. In 2023, this increasing interest led to a highly productive BIRS workshop, organized by the first and third author and Linda Westrick, on this and related topics. We therefore felt the time had come to write a book which would synthesize all that is known to date for anyone seeking an entrance to the discipline. Moreover, we have endeavored to place the foundations of effective metric structure theory on the shoulders of continuous model theory much as classical computable structure theory is founded upon classical model theory.

We have not attempted to write a textbook. There are no exercises. However, this text could be the basis for a course in this area, especially a seminar.

We presume our readers are familiar with first-order logic, but otherwise we have tried to keep the book self contained. In particular, we have included an appendix that summarizes all of the necessary knowledge from computability theory.

We have also included an appendix detailing some of the main open problems in the field as of the time of this writing; we hope it will intrigue newcomers to the area.

We well understand the variety of backgrounds that can lead one to effective metric structure theory. The first author's first research interest was algorithmic randomness. As she progressed through her sequence of postdocs, she became active in computable structure theory and computable analysis, and this combination of interests naturally led her to effective metric structure theory.

The second author is a model theorist by training, but became interested in effective metric structure theory when it played a role in the model-theoretic treatment of the negative resolution of the Connes Embedding Problem in von Neumann algebra theory. Since that project, he has become interested in other applications of effective metric structure theory to operator algebras.

The third author received significant training in computable structure theory from his advisor, Valentina Harizanov, as a graduate student. Afterwards, he made contributions to several areas of computability theory before finding his passion for computable analysis. Upon seeing a talk by A.G. Melnikov about the computable structure theory of metric spaces and L^p spaces, effective metric structure theory became his newfound passion.

Regardless of your background, we hope that you, the reader of this book, will become equally intellectually engaged in this topic.

Hempstead, USA — Johanna N. Y. Franklin
Irvine, USA — Isaac Goldbring
Ames, USA — Timothy H. McNicholl
September 2025

Acknowledgements This book truly began with the workshop *A Convergence of Computable Structure Theory, Analysis, and Randomness* at the Banff International Research Station in March 2023, organized by the first and third authors with Linda Westrick. We would like to thank not only our colleagues, with whom we had somany fruitful discussions there, but also the staff on site who did such an excellent job facilitating the technical aspects of our hybrid workshop and making it possible for the researchers present in Banff to have meaningful conversations with our colleagues in places such as Singapore, New Zealand, Russia, and France—and Ames and Irvine, for two of us were unable to attend the workshop in person. All of these interactions encouraged us, and within a week of the end of the workshop, we had agreed that we would write this book together.

We gratefully acknowledge the funding agencies and the programs that supported our work. The first author was partially supported by a Hofstra University Faculty Research and Development Grant. The second author was partially supported by NSF Grant DMS-2054477. The third author was supported in part by the Scott Hanna Fund of Iowa State University.

We further note the ongoing support of our home institutions. The first author additionally wishes to acknowledge the hospitality of the School of Mathematics and Statistics at Victoria University of Wellington—Te Herenga Waka during her 2024 sabbatical, where she did the majority of her writing.

This book would certainly not exist without the assistance and nudging of the staff at Springer, in particular Elizabeth Loew, who enthusiastically supported this project from the beginning. And while it would exist without the thoughtful comments of our colleagues, who read the drafts of each chapter so carefully, it would not be nearly as good. Therefore, we thank these readers, chief among them Alec Fox, and the anonymous referees.

Finally, we thank our families for their patience with us throughout the long process of making this book a reality.

Competing Interests The authors have no competing interests to declare that are relevant to the content of this manuscript.

Contents

Chapter 1
Introduction

1.1 What Is Effective Metric Structure Theory?

If you are reading these pages now, you may have been drawn to the text by sheer intellectual curiosity about its title. Or, you may already know something about the program of effective metric structure theory and are looking for a source. For those in the former category, we begin with some exegesis on the title.

1.1.1 What Is a Metric Structure?

By a metric structure, we mean any mathematical structure $\mathcal{M}$ that has the following features:

- A set of points, called the *universe* of $\mathcal{M}$, which we denote by M.
- A complete metric on $\mathcal{M}$, which we denote by d_M.
- A (possibly empty) set $\mathrm{Op}_{\mathcal{M}}$ of distinguished operations. Each of these operations is a uniformly continuous map from M^n into M for some positive integer n.
- An associated set of scalars $\mathbb{F}_{\mathcal{M}} \in \{\mathbb{R}, \mathbb{C}\}$.
- A (possibly empty) set $\mathrm{F}_{\mathcal{M}}$ of distinguished functionals. Each of these functionals is a continuous map from M^n to $\mathbb{F}_{\mathcal{M}}$ for some positive integer n.
- A (possibly empty) set $\mathrm{Const}_{\mathcal{M}}$ of distinguished constants. Each of these constants is simply a particular point of M.

Metric structures are ubiquitous in analysis. The simplest example of a metric structure is just a complete metric space where the sets of distinguished operations, functionals, and constants are all empty. Banach spaces may also be thought of as metric structures. Here, the distinguished operations are addition as well as the operations of scalar multiplication $v \mapsto \alpha v$, one such operation for each scalar α. In addition, the distinguished functional is the norm, and the distinguished constant is the zero vector.

J. N. Y. Franklin et al., *Effective Metric Structure Theory*, Springer Monographs in Mathematics, https://doi.org/10.1007/978-3-032-13322-9_1

In a similar way, Hilbert spaces are also natural examples of metric structures, where one adds the inner product as a distinguished functional (in which case one could remove the norm as a distinguished functional as it is recoverable from the inner product). Other interesting metric structures can be obtained from Banach spaces by adding additional operations. Banach algebras and Banach *-algebras are examples of such structures, the latter of which include the important class of C*-algebras.

A perhaps less obvious example of a metric structure comes from probability theory. Indeed, any probability space can be regarded as a metric structure by making the following designations.

- The events of the space (where an event is the equivalence class of a measurable set modulo the relation of null symmetric difference) as the points.
- The metric as the probability of the symmetric difference between events.
- The distinguished operations as those induced by union, intersection, and complementation.
- The probability measure as a distinguished functional.
- The distinguished constants as the events corresponding to the empty event and the whole space.

We note that completeness of the metric is a consequence of σ-additivity of the probability measure.

We finally remark that the class of metric structures is rich enough to include all discrete structures from algebra and combinatorics when one equips these structures with the discrete metric. For example, any field becomes a metric structure when equipped with the discrete metric, the distinguished operations being the addition and multiplication operations, and the distinguished constants being those for the additive and multiplicative identities. An ordered field would also have a distinguished functional for the ordering via its characteristic function. Although our primary interest will be in uncountable metric structures such as those described above, it is worth noting that the theory developed in this book generalizes its classical predecessor in this way.

1.1.2 The Meaning of "Effective"

Within the field of mathematical logic, the term "effective" is synonymous with "computable." Thus, our topic is the theory of computation on metric structures. This is accomplished by merging classical computable structure theory, which pertains to the computability theory of discrete, countable, algebraic structures, with the discipline of computable analysis, which pertains to computation on continuous structures such as the real line.

1.2 Aims of This Book and Intended Audience

We have the following goals for this text:

- To provide a summary of all that is known to date in the area.
- To provide a guide to the various methods used in the field.
- To provide a resource for current scholars in the field as well as newcomers.

In addition, and most importantly, *we will provide a foundation for effective metric structure theory based on continuous logic.*

Our intended audience is anyone who is interested in learning more about the topic and has a good background in at least one of the following areas: functional analysis, model theory, or computability theory. We assume some basic knowledge of first-order logic.

We generally eschew detailed proofs. These can be found in the referenced works, and we intend this work to be a guide to where to look further. We have made exceptions, though, for results that are "well known" but have no complete proof in the literature or for which the proof sheds a particularly interesting light on the concepts in question. We hope these proofs will be especially useful to those seeking an entrance to the subject. We have also provided proof outlines when a proof contains a valuable idea that we believe will be valuable for all scholars in the field.

1.3 Motivations for Studying Effective Metric Structure Theory

A large part of the appeal of studying and developing effective metric structure theory (EMST) is that it draws on results, concepts, and methods from several seemingly disparate areas of mathematics such as computability theory, model theory, functional analysis, and probability theory. Among the most interesting and fruitful connections is the recent confluence of the model theory and computability theory of metric structures. Indeed, techniques from model theory have produced new results in the computability theory of operator algebras and vice versa. We believe the hybridization of these fields will yield many more discoveries. Accordingly, we devote a whole chapter to the model theory of metric structures.

While great progress has been made over the past decade, there are still a number of open problems in EMST, many of which have eluded the usual construction-heavy techniques of computable algebra. Progress is more often obtained by coming to a deep understanding of the related principles of analysis and figuring out how to connect them with the apparatus of computability theory. As we shall see in the subsequent chapters, new results in analysis often must be proven in order to achieve progress on the computability-theoretic side. We believe the solutions of the open problems discussed in this text will result in the development of other such techniques and connections.

All this being said, the beauty EMST achieves though the synthesis of discrete and continuous fields of mathematics is not the sole, and perhaps not even the primary, motivation for its pursuit. It has played a key role in recent solutions of significant problems in analysis. In particular, the degrees of theories of operator algebras play a central role in the model-theoretic approach to the Connes Embedding Problem by Goldbring and Hart [1, 2]. As will be discussed in Chap. 7, Fox, Goldbring, and Hart have recently outlined a program by which EMST might be used to solve the Kirchberg embedding problem in C*-algebra theory.

Finally, to appreciate how we arrive at this point, we now consider the genetics of EMST.

1.4 The Genetics of EMST

The following diagram may help to understand the evolution of EMST.

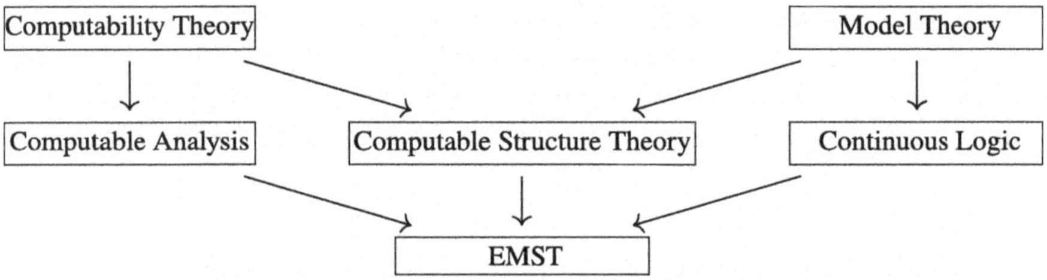

In the remainder of this section, we give a very brief summary of each topic in this diagram and the transitions between them, which are represented by the arrows in the diagram.

1.4.1 *Computability Theory*

Computability theory is built upon four fundamental concepts: partial computable functions, computable sets, computably enumerable sets, and relative computability. These can be defined somewhat informally as follows.

When A, B are sets, we write $f :\subseteq A \to B$ to indicate that f is a function such that $\mathrm{dom}(f) \subseteq A$ and $\mathrm{ran}(f) \subseteq B$. Such a function is called a *partial function from A to B*. Since $\mathbb{N}$ denotes the set of nonnegative integers, a partial function $f :\subseteq \mathbb{N}^n \to \mathbb{N}$ is *computable* if there is an algorithm that, given any $\mathbf{x} \in \mathbb{N}^n$, computes $f(\mathbf{x})$ if $\mathbf{x} \in \mathrm{dom}(f)$ and does not halt if $\mathbf{x} \notin \mathrm{dom}(f)$. A set $A \subseteq \mathbb{N}^n$ is *computable* if its characteristic function χ_A is computable. A set $A \subseteq \mathbb{N}^n$ is *computably enumerable* (c.e.) if it is the domain of a computable partial function. One of the first and most often applied results of computability theory is the existence of a certain c.e. incomputable set, the so-called *halting set*.

Relative computability is somewhat more difficult to define informally. If $A, B \subseteq \mathbb{N}$, we say A is *computable relative to* B if there is an algorithm that, given any $x \in \mathbb{N}$, may ask questions of the form "is $y \in B$?" and, given correct answers to these questions, determines whether $x \in A$. If A is computable relative to B, then we write $A \leq_{\mathrm{T}} B$. It is easily seen that $\leq_{\mathrm{T}}$ is a preorder on the power set of $\mathbb{N}$. If each of A and B is computable relative to the other, then we say that A and B are *Turing equivalent* and write $A \equiv_{\mathrm{T}} B$. The equivalence classes of $\equiv_{\mathrm{T}}$ are known as *Turing degrees*, and the preorder $\leq_{\mathrm{T}}$ induces a partial order on the set of Turing degrees, also denoted by $\leq_{\mathrm{T}}$. A significant part of computability theory is the study of the partial ordering $\leq_{\mathrm{T}}$ on the set of Turing degrees.

We discuss the concepts of computability theory in more detail in Appendix B.

1.4.2 Model Theory

The origins of model theory lie in Tarski's work on the definability of truth. To this end, he assumed that each statement under consideration could be represented as a statement in a *formal language*. Each such language L consists of the following:

- For each positive integer n, a (possibly empty) set of n-ary function symbols and a (possibly empty) set of n-ary relation symbols.
- A (possibly empty) set of constant symbols.

Sentences of L are formed from these symbols by using the equality sign, variables, logical connectives ("$\vee$", "$\wedge$", "$\Rightarrow$", etc.) and quantifiers "$\forall$" and "$\exists$". The truth of such a sentence is undetermined until one specifies an *interpretation* of the language. An interpretation $\mathcal{A}$ begins by specifying a (nonempty) set of objects A, which is called the domain of the interpretation $\mathcal{A}$. One then makes the following associations.

- Each n-ary function symbol f of L is associated with a function $f^{\mathcal{A}} : A^n \to A$.
- Each n-ary relation symbol R of L is associated with a subset $R^{\mathcal{A}} \subseteq A^n$.
- Each constant symbol c of L is associated with an object $c^{\mathcal{A}} \in A$.

A more common name for such an interpretation is a *structure*, or, more precisely, an L-structure. Many familiar mathematical structures, such as groups, rings, and graphs, are structures in this precise sense (for appropriate languages).

An L-*theory* is a set of L-sentences. When T is an L-theory such that each of its sentences are true in some L-structure $\mathcal{A}$, we refer to $\mathcal{A}$ as a *model of* T. Model theory can roughly be described as the study of the classes of models of theories.

All this is to say that structures in and of themselves form a domain of mathematical study quite apart from their philosophical origins. Marker's text provides an excellent introduction to the subject [3]. Model theory has in fact now evolved into a tool that is applied in a wide variety of mathematical disciplines, including number theory, topology, algebra, and data science.

1.4.3 Computable Structure Theory

Computability, as described above, is a theory of computation on the natural numbers. However, we generally conceive of computation as occurring on other domains such as the integers, the rational numbers, groups, graphs, etc. Each of these domains is associated with a collection of operations and relations, that is, each is a structure. Computability is transferred from $\mathbb{N}$ to these structures by means of *presentations*. A *presentation* of a structure $\mathcal{A}$ is a surjection ν from $\mathbb{N}$ onto the domain of $\mathcal{A}$. Note that, for $\mathcal{A}$ to have a presentation, its domain must be countable! If the pullback of the equality relation on the domain of $\mathcal{A}$ under ν along with the pullbacks of the functions and relations of $\mathcal{A}$ are computable, then $(\mathcal{A}, \nu)$ is a *computable presentation*. Computable structure theory is the computability-theoretic study of presentations of structures. We say more about this theory in Chap. 3.

1.4.4 Computable Analysis

Much scientific work relies on computation on continuous domains such as the real line. It may seem, at first glance, that computability theory is incapable of handling this transition since, by Cantor's Theorem, there is no presentation of the real line. However, by consistently and everywhere thinking in terms of approximations, computability theory naturally extends to the real line and to many other continuous domains. For example, we define a real number x to be *computable* if it is possible to compute arbitrarily good approximations to x; that is, there is an algorithm that, given any natural number k as input, produces a rational number q so that $|x - q| < 2^{-k}$. Equivalently, x is computable if there exists an algorithm for computing the base 10 (or any number base) expansion of x.

Computability of a function $f : \mathbb{R} \to \mathbb{R}$ can also be sensibly defined by means of approximations; namely, f is computable if there is an algorithm that, for every real number x, can compute arbitrarily good approximations of $f(x)$ from sufficiently good approximations of x. We will state this idea more formally in Chap. 3.

In short, computable analysis is the study of computability in continuous domains by means of approximations.

1.4.5 Continuous Logic

Just as classical computability theory requires some adjustment when working in continuous settings, so does the model theory of Tarski. In particular, the model theory of Tarski is not up to the task of discussing ultraproducts of metric structures (see the discussion in Chap. 2). The model theory for metric structures uses a variant of classical logic aptly named *continuous logic*. We will discuss this logic in some detail in Chap. 2. The essential ideas are the following:

- Replace the standard set of truth values with the numbers in a closed bounded interval. Usually this set is taken to be $[0, 1]$, with 0 representing "true" and 1 representing "false."
- Replace the equality sign with the distance sign d.
- Relations now become uniformly continuous functions that map into $[0, 1]$. Functions are still functions, but must be uniformly continuous.
- The quantifier $\forall$ is replaced with "sup" and the quantifier $\exists$ is replaced by "inf".

It should be noted that the resulting theory is an extension of the classical model theory, as can be seen by choosing the metric to be the discrete metric.

1.4.6 Effective Metric Structure Theory

Effective metric structure theory ties all of these domains together. As we have discussed, computable structure theory was originally conceived as applying only to countable structures. In 2013, Nies and Melnikov pointed out that the tools of computable analysis could be used to extend computable structure theory to separable uncountable metric structures [4, 5]. Some of their ideas were implicit in the text of Pour-El and Richards [6]. Recently, results from continuous logic have been used to prove results in EMST (see, for example, [7]). Conversely, some results from EMST were used in the model-theoretic refutation of the Connes Embedding Problem due to Goldbring and Hart [1, 2]; for more on this, see Chap. 8.

We emphasize that metric structures are required to have a metric. We do not treat topological structures such as topological groups in this work.

We note also that there is a considerable body of work in computable analysis, particularly on Banach spaces, that is not cited here simply because it does not fit into any of the themes of this book, namely, computable presentability, computable categoricity, degrees of theories, and index sets. This book should not be thought of, and it does not purport to be, an encyclopedic treatment of computable analysis. We refer the reader who wishes to know more about this broad topic to [8].

1.5 The Organization of the Book

We first cover the model theory and computability theory of metric structures in Chaps. 2 and 3. We then attend to the computable structure theory of metric spaces in Chap. 4. Results on Banach spaces are spread out over two chapters. Chapter 5 attends to Banach spaces of continuous functions on a compact space, while Chap. 6 discusses results on Lebesgue spaces. The results on operator algebras are covered in Chap. 7. Chapter 8 deals with degrees of theories.

Along the way, we will pose many open questions that are motivated by the topics at hand; these questions, together with others that did not fit naturally into the text, will be collected in Appendix A. The reader unfamiliar with basic computability theory can consult Appendix B, which provides some background in basic computability theory together with assorted advanced topics needed in the book (such as Kleene's $\mathcal{O}$).

1.6 A Few Matters of Notation and Terminology

We denote the set of natural numbers (that is, the set of nonnegative integers) by $\mathbb{N}$. We use $\mathbb{F}$ to denote our field of scalars, which is always $\mathbb{R}$ or $\mathbb{C}$. When A is a set, χ_A denotes its characteristic function. That is,

$$\chi_A(x) = \begin{cases} 1 & x \in A \\ 0 & x \notin A \end{cases}.$$

The domain of χ_A is always made clear by context.

If A is a set, then $A^* = \bigcup_{n\in\mathbb{N}} A^n$. In other words, A^* is the set of all finite sequences whose terms are all elements of A; note that this includes the empty sequence. A *tree* is a nonempty subset of $\mathbb{N}^*$ that is closed under prefixes. Thus, every tree contains the empty sequence.

If A is a set in a topological space, we write $\overline{A}$ for the closure of A and A° for the interior of A.

If $\mathcal{V}$ is a vector space, then we write $\mathbf{0}_{\mathcal{V}}$ for the additive identity of $\mathcal{V}$. When our context makes our intention clear, we simply write $\mathbf{0}$. We also use $\mathbf{0}$ for the Turing degree of the computable sets and trust context to make the required distinctions.

References

1. Isaac Goldbring and Bradd Hart. Computability and the Connes embedding problem. *Bulletin of Symbolic Logic*, 22(2):238–248, 2016.
2. Isaac Goldbring and Bradd Hart. The universal theory of the hyperfinite II_1 factor is not computable. *Bull. Symb. Log.*, 30(2):181–198, 2024.
3. David Marker. *Model Theory: An Introduction*, volume 217 of *Graduate Texts in Mathematics*. Springer-Verlag, New York, 2002.
4. Alexander G. Melnikov. Computably isometric spaces. *J. Symbolic Logic*, 78(4):1055–1085, 2013.
5. Alexander G. Melnikov and André Nies. The classification problem for compact computable metric spaces. In *The Nature of Computation*, volume 7921 of *Lecture Notes in Comput. Sci.*, pages 320–328. Springer, Heidelberg, 2013.

6. Marian B. Pour-El and J. Ian Richards. *Computability in Analysis and Physics*. Perspectives in Mathematical Logic. Springer-Verlag, Berlin, 1989.
7. Alec Fox. Computable presentations of C*-algebras. *J. Symb. Log.*, 89(3):1313–1338, 2024.
8. Vasco Brattka and Peter Hertling, editors. *Handbook of Computability and Complexity in Analysis*. Theory and Applications of Computability. Springer, Cham, 2021.

Chapter 2
An Introduction to Metric Structures and Their Model Theory

In this chapter, we introduce the main objects of study throughout this book, namely metric structures, as well as the associated continuous logic suitable for analyzing them from a logical perspective. The reader interested in further references for the material presented in this chapter may consult [1, 2]. We intend to use this logic as a foundation for studying the computability of these structures. In Chaps. 7 and 8, we will see the interplay between the model theory and computability theory of metric structures.

2.1 A Motivation: Metric Ultraproducts

While the focus of this book is on *effective* matters, we begin our discussion of the mechanics of the model theory of metric structures by considering an essential, but *in*effective construction, namely the ultraproduct construction. Just as ultraproducts are used to characterize first-order logic, they are also used to characterize continuous logic.

Consider a family $(M_n, d_n)_{n\in\mathbb{N}}$ of metric spaces where each space has diameter at most 1.[1] Fix an ultrafilter $\mathcal{U}$ on $\mathbb{N}$.[2] Define a pseudometric d on $\prod_n M_n$ by setting $d(a, b) = \lim_{\mathcal{U}} d_n(a_n, b_n)$.[3] As with any pseudometric, we should consider the equivalence relation on $\prod_n M_n$ given by the relation $d(a, b) = 0$. We denote the equivalence class of $a \in \prod_n M_n$ by $a_{\mathcal{U}}$. The collection of these equivalence classes

[1] There is nothing especially important about the bound 1; all that is germane to the discussion that follows is that there is some uniform upper bound on the diameters of the spaces in the family.

[2] Recall that an ultrafilter $\mathcal{U}$ on a set I is a finitely-additive $\{0, 1\}$-valued probability measure on all subsets of I; we usually abuse notation and identify an ultrafilter with its sets of measure 1.

[3] For any bounded sequence $(r_i)_{i\in\mathbb{I}}$ of real numbers and ultrafilter $\mathcal{U}$ on I, $\lim_{\mathcal{U}} r_i$ is defined to be the unique real number r with the property that, for any $\epsilon > 0$, one has $|r - r_i| < \epsilon$ for $\mathcal{U}$-almost all $i \in \mathbb{I}$ and is called the $\mathcal{U}$-ultralimit of the sequence $(r_i)_{i\in I}$.

J. N. Y. Franklin et al., *Effective Metric Structure Theory*, Springer Monographs in Mathematics, https://doi.org/10.1007/978-3-032-13322-9_2

is called the *(metric) ultraproduct* of the family $(M_n, d_n)_{n\in\mathbb{N}}$ with respect to the ultrafilter $\mathcal{U}$ and denoted by $\prod_{\mathcal{U}}(M_n, d_n)$ or sometimes simply $\prod_{\mathcal{U}} M_n$. If each (M_n, d_n) equals a fixed metric space (M, d), then we speak of the *ultrapower* $(M, d)^{\mathcal{U}}$ of (M, d) with respect to $\mathcal{U}$, which may often be shortened to simply $M^{\mathcal{U}}$. In this case, there is a natural *diagonal embedding* $M \hookrightarrow M^{\mathcal{U}}$ given by mapping the element a to the equivalence class of the sequence which is constantly a. Note that the diagonal embedding is an isometric inclusion.

Now suppose that we fix a distinguished function $f : M \to M$. Can one "extend" f to $M^{\mathcal{U}}$ in the obvious pointwise fashion, that is, by defining $f(a_{\mathcal{U}}) = f(a)_{\mathcal{U}}$? Here, $f(a)$ denotes the sequence $n \mapsto f(a_n)$. Note that this extension is possible if f is uniformly continuous. Indeed, the question of well-definedness of the extension boils down to the question of whether or not $d(a, b) = 0$ implies $d(f(a), f(b)) = 0$. To see that this is the case, fix $\epsilon > 0$. We show that $d(f(a), f(b)) \le \epsilon$. Choose $\delta > 0$ so that for all $x, y \in M$, if $d(x, y) < \delta$, then $d(f(x), f(y)) \le \epsilon$.[4] Since $d(a, b) = 0$, we know that $d(a_n, b_n) < \delta$ for $\mathcal{U}$-almost all $n \in \mathbb{N}$. Whence, $d(f(a_n), f(b_n)) \le \epsilon$ for this same set of $n \in \mathbb{N}$, and thus $d(f(a), f(b)) \le \epsilon$ as desired. Notice also that a similar argument shows that the extension of f to $M^{\mathcal{U}}$ is once again uniformly continuous as witnessed by the same *modulus of uniform continuity* $\epsilon \mapsto \delta(\epsilon)$.[5]

Conversely, uniform continuity is in fact necessary for this extension to be possible. To see this, suppose that f is not uniformly continuous. Then, there exist $\epsilon > 0$ and sequences $(a_n)_{n\in\mathbb{N}}$ and $(b_n)_{n\in\mathbb{N}}$ from M so that $d(a_n, b_n) < \frac{1}{n+1}$ and yet $d(f(a_n), f(b_n)) \ge \epsilon$. Presuming that $\mathcal{U}$ is nonprincipal (that is, is not a Dirac measure, so cofinite sets have measure 1), we have that $d(a, b) = 0$ and yet $d(f(a), f(b)) \ge \epsilon$.

A similar analysis applies to the more general case of an ultraproduct $\prod_{\mathcal{U}} M_n$ of a family of potentially different metric spaces, each equipped with a distinguished function f_n. To begin, define $f(a)$ to be the sequence $(f_n(a_n))_{n\in\mathbb{N}}$. One can argue as in the previous paragraph that if the sequence $(f_n)_{n\in\mathbb{N}}$ is uniformly equicontinuous, then the map $a \mapsto f(a)$ is once again well defined and uniformly continuous with the same modulus of uniform continuity as the individual functions f_n. Note also that the same idea can be used to treat multivariate functions $f_n : M_n^m \to M_n$ with an appropriate choice of metric on M_n^m (say, for example, the maximum metric).

In classical logic, one not only considers distinguished functions but also distinguished predicates (or relations), which are viewed as distinguished subsets of (cartesian powers of) the universe of the structure. What should be the analog in the metric situation? Well, in classical logic, the equality symbol is a privileged relation, so it seems fitting that the metric should play the analogous role in our context. Viewing equality as a map into $\{0, 1\}$ and similarly the metric as a map into $[0, 1]$ (recall our bounded diameter assumption), it seems that distinguished predicates should take the form $P : M \to [0, 1]$ (or really any compact interval in $\mathbb{R}$). One can then take the ultraproduct of the structures (M_n, d_n, P_n) as above, provided that the P_n's

[4] The reason for the asymmetry in the inequalities will become clear in a moment.

[5] This statement would not be true if only strong inequalities were used in the definition of modulus of uniform continuity.

are uniformly equicontinuous and we use the usual metric on [0, 1]. At first glance this might seem slightly unsatisfactory as the resulting distinguished predicate P on $\prod_{\mathcal{U}} M_n$ is given by $P(a_{\mathcal{U}}) = (P_n(a_n))_{\mathcal{U}} \in [0, 1]^{\mathcal{U}}$. However, using the canonical isometry $r_{\mathcal{U}} \mapsto \lim_{\mathcal{U}} r_n$ between $[0, 1]^{\mathcal{U}}$ and [0, 1], we can safely view this distinguished predicate as taking values in [0, 1]. We note this is where the compactness of [0, 1] proves crucial. Once again, P will be uniformly continuous with the same modulus of uniform continuity as the individual P_n's.

2.2 Syntax and Semantics

2.2.1 Basic Definitions

Given the discussion in the previous section, it should be fairly unsurprising how to define continuous languages (or signatures) and structures in those languages. Indeed, the key data to keep track of were the diameters of the spaces involved, the moduli of uniform continuity of the various distinguished functions and predicates, and the compact intervals where the predicates took their values. We formalize this into the following definition.

Definition 2.1 A *(continuous) signature* (or *continuous language*) L consists of the following:

1. A bound $D_L \geq 0$.
2. A set $\mathcal{F}$ of *function symbols*. For each $f \in \mathcal{F}$, the signature provides
    a. an arity $n_f \in \mathbb{N}$, and
    b. a *modulus of uniform continuity* $\Delta_f : (0, 1) \to (0, 1)$.
3. A set $\mathcal{R}$ of *predicate symbols*. For each $P \in \mathcal{R}$, the signature L provides
    a. an arity $n_P \in \mathbb{N}$,
    b. a *modulus of uniform continuity* $\Delta_P : (0, 1) \to (0, 1)$, and
    c. a bound $R_P \geq 0$.

Remark 2.2

1. We always view the metric symbol as a distinguished predicate symbol, just as classical logic views the equality sign as a distinguished predicate symbol. The bound provided in the third requirement for predicate symbols should coincide with the bound D_L provided by the signature.
2. As in classical logic, we can consider the case of function symbols with arity 0, that is, constant symbols which will serve to isolate distinguished elements of our structure. In this case, there is no modulus of uniform continuity requirement.

3. It is straightforward to extend the previous definition to the *many-sorted* situation, that is, to the situation where the structure may have more than one sort (or universe). In this case, the signature needs to provide a bound for each sort. In addition, function and predicate symbols take inputs from some finite cartesian product of the sorts, and predicates output values in some sort. The many-sorted setting will become important when we want to treat certain *unbounded* metric structures.

Equipped with the notion of a continuous signature, we can now provide the definition of a structure in such a signature.

Definition 2.3 Given a continuous signature L, an L-*structure* $\mathcal{M}$ specifies the following information:

1. A *complete metric space* (M, d) whose diameter is bounded by D_L.
2. For each function symbol f, a function $f^{\mathcal{M}} : M^{n_f} \to M$ for which Δ_f is a modulus of continuity.
3. For each predicate symbol P, a function $P^{\mathcal{M}} : M^{n_P} \to [-R_P, R_P]$ for which Δ_P is a modulus of continuity.

In the first item in the above definition, we emphasized that the underlying universe of a metric structure must be a complete metric space. At first glance, this might seem like a stringent requirement. Indeed, in analysis one often encounters incomplete metric spaces and even mere pseudometric spaces. From the model-theoretic perspective, however, the requirement of a complete metric space is rather mild. Indeed, to any so-called L-*prestructure* based on a pseudometric space, one can always first pass to the associated metric space (obtained by identifying points at distance 0 from each other) and then take the completion in such a way that the resulting complete metric space carries natural interpretations of the symbols and for which the semantics of the original L-prestructure and the resulting completed L-structure are identical; see [1, Sect. 3] for details.

It is straightforward to extend the notion of an L-structure to the case that L is a many-sorted language as described in Remark 2.2(3) above.

2.2.2 Examples

We now describe a handful of examples that will be treated in depth throughout this book.

Example 2.4 One can consider a bounded complete metric space (M, d) with no further structure, that is, with no distinguished function or predicate symbols. Although seemingly trivial, the effective structure theory of such spaces is already quite interesting, as will be discussed in Chap. 4.

Example 2.5 For $\mathbb{F} \in \{\mathbb{R}, \mathbb{C}\}$, recall that a *Banach space over* $\mathbb{F}$ is a vector space over $\mathbb{F}$ equipped with a norm whose associated metric is complete. We would like to treat normed vector spaces as metric structures. However, an issue arises immediately: *the metric on a Banach space is unbounded!*

There are two competing remedies for this issue. The first idea is to note that all of the interesting geometry of a Banach space is already apparent in the unit ball of the Banach space (which is, of course, bounded). Thus we may take the unit ball of the Banach space, equipped with its usual metric, to be the underlying universe of the structure. In this case, the language provides a bound of 2 on the metric. The signature should include symbols to reflect the linear structure of the Banach space. However, to ensure that the values of the function symbols remain in the unit ball, one must include binary function symbols for "rounded combinations" $(x, y) \mapsto \alpha x + \beta y$ for $(\alpha, \beta) \in \mathbb{F}^2$ with $|\alpha| + |\beta| \leq 1$. One also includes a constant symbol for the zero vector $\mathbf{0}$ and a unary predicate symbol $\|\cdot\|$ for the norm on the Banach space with bound 1 (although including the norm as a distinguished predicate is technically unnecessary as it is recoverable from the metric symbol, which is always included, and the zero vector). The moduli of uniform continuity for the aforementioned binary function symbols and unary predicate symbol for the norm are the natural ones.

The other approach is to work in a many-sorted language for Banach spaces consisting of a countable family of sorts $(B_n)_{n\in\mathbb{N}}$ whose natural interpretations are the open balls of radius n centered around the origin. In this case, the language provides a bound of $2n$ on the sort B_n. We now have a host of function symbols for the linear structure, taking into account where elements from each sort might land under the various operations. In particular, one has binary function symbols $+_n : B_n \times B_n \to B_{2n}$. As an example of the symbols involved for scalar multiplication, for each $n \in \mathbb{N}$, there is a unary function symbol $4\cdot : B_n \to B_{4n}$ for multiplication by the scalar 4. One also has distinguished constant symbols $\mathbf{0}_n \in B_n$, all naming the zero vector, and distinguished predicate symbols $\|\cdot\|_n$ with domain B_n and bound n. Finally, to relate the sorts, one provides unary function symbols $\iota_{m,n} : B_m \to B_n$ for inclusion mappings whenever $m < n$. All of these symbols should be provided with their natural moduli of uniform continuity.

In either case, it is clear how to view a Banach space over $\mathbb{F}$ as a structure in the appropriate language. In the latter approach, we sometimes call the associated structure the *dissection* of the Banach space (as we have chopped up the poor Banach space into these various sorts in order to take apart its natural linear structure and replace it with this entire array of distinguished function and predicate symbols).

Remark 2.6 One may note that since the field $\mathbb{F}$ is uncountable, either language for studying Banach spaces is uncountable. However, for nearly every purpose, restricting the symbols for scalar multiplication to include only the rational scalars (that is, $\mathbb{Q}(i) \cap \mathbb{F}$) creates no loss in expressive power. Consequently, we may (and usually will) treat the language for Banach spaces as a countable language. The same remark holds for the examples that follow.

Example 2.7 For $\mathbb{F} \in \{\mathbb{R}, \mathbb{C}\}$, recall that a *Hilbert space over* $\mathbb{F}$ is an inner product space over $\mathbb{F}$ whose associated metric is complete. Once again, the metric is unbounded and so we can choose to treat Hilbert spaces in one of the two approaches presented in the previous example. Since Hilbert spaces have more structure than Banach spaces, we should reflect this in our choice of language by including symbols for the inner product. In the case of real Hilbert spaces, in the unit ball approach, we should add a binary predicate symbol $\langle \cdot, \cdot \rangle$ for the inner product with bound 1 while in the many-sorted approach, we should include one such symbol for each n with bound n^2. If the Hilbert space is complex, then the inner product can take complex values while we have insisted that our predicates take real values. Consequently, we must include two copies of the above predicate symbols, one for the real part of the inner product and one for the imaginary part. In all of these cases, the moduli of uniform continuity are the natural ones.

Example 2.8 One can also augment Banach spaces by adding extra algebraic structure (as opposed to adding an inner product as in the previous example). For example, when considering a *Banach algebra*, which is a Banach space equipped with a multiplication operation that interacts nicely with the vector and norm structure (see Chap. 7 for a precise definition), one would add a distinguished binary function symbol for the multiplication (or one such symbol for each sort in the many-sorted approach) with the obvious modulus of uniform continuity. If the Banach algebra is *unital* (that is, has a multiplicative unit), then one may include a constant symbol (or infinitely many such constant symbols) to name the unit. If the Banach algebra is in fact a *Banach* $*$*-algebra*, that is, carries an involution $*$ that interacts nicely with the Banach algebra structure, then one would add a unary function symbol (or infinitely many such symbols) to name the involution. Unital Banach $*$-algebras would consequently name the multiplication symbol, the involution, and the unit. Important examples of Banach $*$-algebras are C^**-algebras*, whose effective structure theory will be discussed at length in Chaps. 7 and 8.

2.2.3 Formulas and Their Interpretation

Throughout this subsection, we fix a continuous signature L. One defines L-terms as in classical logic, namely, they are obtained from variables and constant symbols using function symbols. Given the definition of L-terms, we can now define L-formulas.

Definition 2.9 The set of L*-formulas* is defined by recursion on their complexity as follows:

1. If P is a predicate symbol in L and $t_1, \ldots, t_{n_P}$ are L-terms, then $P(t_1, \ldots, t_{n_P})$ is an L-formula.
2. If $\varphi_1, \ldots, \varphi_n$ are L-formulas and $u : \mathbb{R}^n \to \mathbb{R}$ is a continuous function, then $u(\varphi_1, \ldots, \varphi_n)$ is an L-formula.
3. If φ is an L-formula and x is a variable, then $\sup_x \varphi$ and $\inf_x \varphi$ are L-formulas.

The formulas given in the first clause of Definition 2.9 are called *atomic L-formulas*. Note that, since the metric symbol d is always considered a distinguished predicate symbol in any continuous language, one always has the atomic L-formulas $d(t_1, t_2)$.

The continuous functions in the second clause of Definition 2.9 are often referred to as *connectives* and are the continuous analog of the usual Boolean connectives, which are simply functions from $\{0, 1\}^n \to \{0, 1\}$. Just as one can generate any Boolean function using a very small set of Boolean functions (in fact, a single Boolean function suffices), one can use a small set of connectives that generates a dense set of L-formulas. Indeed, by [1, Sect. 6], the formulas formed using only the unary connectives $x \mapsto 0$, $x \mapsto 1$, and $x \mapsto \frac{x}{2}$ and the binary connective $(x, y) \mapsto x \dot{-} y := \max(x - y, 0)$, called the *restricted L-formulas*, are dense in the set of all formulas with respect to the pseudometric on L-formulas defined in Definition 2.12 below. Consequently, if L is a countable, "effectively enumerable" language (to be defined precisely in Chap. 3), then the set of restricted L-formulas is also effectively enumerable and is, moreover, dense in the set of all L-formulas.

In the third clause in Definition 2.9, we view sup and inf as quantifiers which play a role somewhat analogous to the usual first-order quantifiers $\forall$ and $\exists$. Shortly, we will see the reason behind this analogy. Given this discussion, it makes sense to define *free* and *bound* occurrences of variables just as in classical logic.

As in classical logic, given an L-formula $\varphi(\mathbf{x})$,[6] an L-structure $\mathcal{M}$, and a tuple $\mathbf{a} \in M$, one would like to interpret the formula φ in $\mathcal{M}$ upon substituting the tuple $\mathbf{a}$. As a simple example, if $\varphi(\mathbf{x})$ is the atomic formula $d(x_1, x_2)$, then the interpretation of this formula in $\mathcal{M}$ upon substituting $\mathbf{a}$ should be the distance in M between a_1 and a_2, which is some real number. We thus have an important difference between classical logic and continuous logic: in the latter, interpretations of formulas are real numbers. We now give the general definition. As in classical logic, one needs to define the interpretation of terms.

Definition 2.10 Given an L-term $t(\mathbf{x})$, an L-structure $\mathcal{M}$, and a tuple $\mathbf{a} \in M$, we define the *interpretation* $t^{\mathcal{M}}(\mathbf{a})$ by recursion on the complexity of t as follows.

1. If $t(\mathbf{x})$ is a variable x_i, then $t^{\mathcal{M}}(\mathbf{a}) = a_i$.
2. If $t(\mathbf{x})$ is a constant symbol c, then $t^{\mathcal{M}}(\mathbf{a}) = c^{\mathcal{M}}$.
3. If $t(\mathbf{x}) = f(t_1(\mathbf{x}), \ldots, t_{n_f}(\mathbf{x}))$ where each t_i is a term and f is a function symbol, then
$$t^{\mathcal{M}}(\mathbf{a}) = f^{\mathcal{M}}\left(t_1^{\mathcal{M}}(\mathbf{a}), \ldots, t_{n_f}^{\mathcal{M}}(\mathbf{a})\right).$$

The reader can verify that each function $t^{\mathcal{M}} : M^n \to M$ (where n is the number of variables appearing in t) is uniformly continuous and uniformly so over all L-structures. In other words, there is a function $\Delta_t : (0, 1) \to (0, 1)$ such that for every L-structure $\mathcal{M}$, Δ_t is a modulus of continuity for $t^{\mathcal{M}}$. We sometimes refer to this fact

[6] As usual, the notation indicates that the free variables of φ are among those occurring in $\mathbf{x}$.

by saying that the term has a modulus of uniform continuity "known to the signature L."

We are now ready to define the interpretation of formulas.

Definition 2.11 Given an L-formula $\varphi(\mathbf{x})$, an L-structure $\mathcal{M}$, and a tuple $\mathbf{a} \in M$, we define the *interpretation* $\varphi^{\mathcal{M}}(\mathbf{a})$ by recursion on the complexity of φ as follows.

1. If $\varphi(\mathbf{x})$ is the atomic formula $P(t_1(\mathbf{x}), \ldots, t_{n_P}(\mathbf{x}))$, then

$$\varphi^{\mathcal{M}}(\mathbf{a}) = P^{\mathcal{M}}\left(t_1^{\mathcal{M}}(\mathbf{a}), \ldots, t_{n_P}^{\mathcal{M}}(\mathbf{a})\right).$$

2. If $\varphi(\mathbf{x}) = u(\varphi_1(\mathbf{x}), \ldots, \varphi_m(\mathbf{x}))$ where each $\varphi_i(\mathbf{x})$ is a formula, then

$$\varphi^{\mathcal{M}}(\mathbf{a}) = u\left(\varphi_1^{\mathcal{M}}(\mathbf{a}), \ldots, \varphi_m^{\mathcal{M}}(\mathbf{a})\right).$$

3. If $\varphi(\mathbf{x}) = \sup_y \psi(\mathbf{x}, y)$, then $\varphi^{\mathcal{M}}(\mathbf{a}) = \sup\{\psi^{\mathcal{M}}(\mathbf{a}, b) : b \in M\}$. The identical definition is made for the case of $\varphi(\mathbf{x}) = \inf_y \psi(\mathbf{x}, y)$.

The reader might wonder why the suprema and infima appearing in the last clause of the previous definition are finite. The reason is that one can prove, by induction on the complexity of formulas, that each $\varphi^{\mathcal{M}}$ is uniformly continuous and bounded. Furthermore, both the modulus of uniform continuity Δ_φ and bound R_φ are known to the language.

As remarked above, in continuous logic, the interpretation $\varphi^{\mathcal{M}}(\mathbf{a})$ is not simply true or false, but rather is a real number (in the interval $[-R_\varphi, R_\varphi]$). If one prefers, one can think of this value as a "truth value" that is somewhere between true and false, but one should not get emotionally attached to such a view. In any case, to make an assertion one needs to set a formula equal to a number. For example, if $\varphi(\mathbf{x}) = d(t_1(\mathbf{x}), t_2(\mathbf{x}))$, then $\varphi^{\mathcal{M}}(\mathbf{a}) = 0$ asserts that the elements $t_1^{\mathcal{M}}(\mathbf{a})$ and $t_2^{\mathcal{M}}(\mathbf{a})$ are identical. For this reason, one often thinks of 0 as "true" and writes $\mathcal{M} \models \varphi(\mathbf{a})$ (read: $\mathcal{M}$ *models* $\varphi(\mathbf{a})$) if $\varphi^{\mathcal{M}}(\mathbf{a}) = 0$. However, it is important to point out that 0 is simply an arbitrary choice and that one can recover other values of formulas from zerosets of formulas by noting that $\varphi^{\mathcal{M}}(\mathbf{a}) = r$ if and only if $\mathcal{M} \models |\varphi(\mathbf{a}) - r|$.[7] We can even use zerosets to express weak inequalities like $\varphi^{\mathcal{M}}(\mathbf{a}) \leq r$ by writing $\mathcal{M} \models \varphi(\mathbf{a}) \dot{-} r$.

Definition 2.12 Given a finite tuple $\mathbf{x}$ of variables, we define a pseudometric on the set $\mathrm{Form}_{\mathbf{x}}(L)$ of all L-formulas with free variables among $\mathbf{x}$ by setting

$$d(\varphi, \psi) = \sup\left\{|\varphi^{\mathcal{M}}(\mathbf{a}) - \psi^{\mathcal{M}}(\mathbf{a})| : \mathcal{M} \text{ an } L\text{-structure}, \mathbf{a} \in M^{\mathbf{x}}\right\}.$$

Here, $M^{\mathbf{x}}$ is the set of all tuples from M appropriate to be substituted for $\mathbf{x}$.

[7] The function $x \mapsto |x - r| : \mathbb{R} \to \mathbb{R}$ is continuous and thus a connective that we can apply to the formula $\varphi(\mathbf{x})$ in order to obtain another formula.

We note that formulas are logically equivalent if and only if their distance with respect to this pseudometric is 0.

Recall that the density character $\mathfrak{c}(X)$ of a topological space X is the smallest cardinality of a dense subset. Accordingly, we also define the *density character* $\mathfrak{c}(L)$ of L to be $\sum_{\mathbf{x}} \mathfrak{c}(\mathrm{Form}_{\mathbf{x}}(L))$. If $\mathfrak{c}(L) = \aleph_0$, we say that L is *separable*.

At this point, we can return to our somewhat cryptic remarks concerning the viewpoint that sup and inf behave like the quantifiers $\forall$ and $\exists$. Indeed, suppose that $\varphi(x)$ is a formula that takes only nonnegative values in all L-structures. Then $\mathcal{M} \models \sup_x \varphi(x)$ if and only if $\mathcal{M} \models \varphi(a)$ for all $a \in M$. Thus, sup does indeed behave like a universal quantifier. On the other hand, $\mathcal{M} \models \inf_x \varphi(x)$ if and only if, for every $\epsilon > 0$, there is an $a \in M$ such that $\mathcal{M} \models \varphi(a) \dot{-} \epsilon$. In this way, inf only behaves like an approximate existential quantifier.[8]

An *L-sentence* is an L-formula σ without free variables. Consequently, in any L-structure $\mathcal{M}$, $\sigma^{\mathcal{M}}$ is some real number, and $\mathcal{M} \models \sigma$ is another way of denoting the fact that $\sigma^{\mathcal{M}} = 0$. An *L-theory* is a set T of L-sentences. An L-structure $\mathcal{M}$ is said to be a *model* of the L-theory T, denoted $\mathcal{M} \models T$, if $\mathcal{M} \models \sigma$ for all $\sigma \in T$. The L-theory T is *satisfiable* if it has a model. Given an L-structure $\mathcal{M}$, the *complete theory of* $\mathcal{M}$ is the L-theory $\mathrm{Th}(\mathcal{M})$, which is defined to be $\{\sigma : \mathcal{M} \models \sigma\}$.[9] We also write $T \models \sigma$ if all models of T also model σ. Two L-theories T_1 and T_2 are *equivalent* if they have the same set of models, and an L-theory T is *complete* if it is equivalent to $\mathrm{Th}(\mathcal{M})$ for some L-structure $\mathcal{M}$.

2.2.4 Thinking Positively

An important aspect of continuous logic to remain aware of is that it is a *positive* logic in that it does not contain a negation symbol. However, there are some tricks for approximately expressing negations that are useful, especially in the context of implications. In classical logic, an implication implicitly uses a negation since $P \Rightarrow Q$ is shorthand for $(\neg P) \vee Q$. One substitute for this in continuous logic is to use the connective $(x, y) \mapsto \min(x, y)$. For example, one can express the implication $(\varphi < \delta) \Rightarrow (\psi \leq \epsilon)$ as $\min(\delta \dot{-} \varphi, \psi \dot{-} \epsilon) = 0$.

Another useful trick is the ability to express implications of the form $\varphi^{\mathcal{M}}(\mathbf{a}) = 0 \Rightarrow \psi^{\mathcal{M}}(\mathbf{a}) = 0$ as $\mathcal{M}$ ranges over all models of some theory T and $\mathbf{a}$ ranges over M. This can be accomplished as follows [1, Sect. 7].

---

[8] In certain "saturated" structures, such as nonprincipal ultraproducts, these approximate existential quantifiers become exact.

[9] An alternate description of $\mathrm{Th}(\mathcal{M})$ is as the function $\sigma \mapsto \sigma^{\mathcal{M}}$ defined on the set of all L-sentences. Given the above remarks, the two descriptions of the complete theory of $\mathcal{M}$ carry the same information. However, in Chap. 8, we will see there is a subtle difference between the two approaches when effective matters are taken into consideration.

Theorem 2.13 *Given an L-theory T and L-formulas $\varphi(\mathbf{x})$ and $\psi(\mathbf{x})$, the following are equivalent.*

1. *For all $\mathcal{M} \models T$ and $\mathbf{a} \in M^{\mathbf{x}}$, $(\varphi^{\mathcal{M}}(\mathbf{a}) = 0) \Rightarrow (\psi^{\mathcal{M}}(\mathbf{a}) = 0)$.*
2. *For all $\epsilon > 0$, there is $\delta > 0$ such that*

$$T \models \sup_{\mathbf{x}} \min(\delta \dot{-} \varphi(\mathbf{x}), \psi(\mathbf{x}) \dot{-} \epsilon).$$

3. *There is an increasing continuous function $\alpha : [0, 1] \to [0, 1]$ such that $\alpha(0) = 0$ and $T \models \sup_{\mathbf{x}}(\psi(\mathbf{x}) \dot{-} \alpha(\varphi(\mathbf{x})))$.*

The power of this theorem is that one can express the implication (1) using the official L-sentence appearing in (3). Of particular importance in connection to effective matters is the fact that if the dependence of δ on ϵ in item (2) is effective, then one can find a "computable" such function α; for more on this, see [3, Sect. 2].

2.2.5 *Axiomatizing Classes of L-Structures*

Earlier, we described several languages suitable for studying structures from analysis. For example, we described a many-sorted language L_B for studying (say, real) Banach spaces. However, an arbitrary L_B-structure need not be (the dissection of) a Banach space, for it merely needs to interpret the symbols and does not have to do so in any sensible manner. Thus, it is important to know that the intended L_B-structures (namely the dissections of Banach spaces) can indeed be captured in some appropriate sense. The precise formulation of this is to say that the class of Banach spaces (viewed as L_B-structures) is an *axiomatizable* or *elementary* class, in the sense that there is some L_B-theory T_B whose models are precisely the (dissections of) Banach spaces. It is straightforward (albeit more than a bit tedious) to describe such an L_B-theory.

For example, one will need axioms to describe the vector space structure of the Banach space. Given the many-sorted nature of our language, we will need many such axioms. For example, to express the fact that multiplication by 4 distributes over addition, we will need an "axiom" of the form

$$\sup_{x,y \in B_n} \|4 \cdot_{2n} (x_n +_n y_n) - ((4 \cdot_n x_n) +_{4n} (4 \cdot_n y_n))\|_{8n},$$

one such axiom for every n. Note that setting this sentence to 0 does indeed tell us that, for elements of the n-ball, multiplication by 4 distributes over addition.[10] We have been pedantic and have labeled all of the various symbols with numbers indicating which sorts they are being applied to. For example, since x_n and y_n belong to B_n, they must be added according to the symbol $+_n$, and the resulting sum belongs

[10] Such a statement is a universal statement and, as described above, our sup quantifier is acting as a universal quantifier.

in the sort B_{2n}. Therefore, we must be using the symbol $4\cdot_{2n}$ for multiplication by 4 on elements in the sort $4n$. The resulting product belongs to the sort $8n$, as does the other term appearing in the difference. The attentive reader may have noticed that the subtraction sign does not have a decoration; given that our language technically does not have a symbol for subtraction, we might mean $+_{8n}$ with a $(-1)\cdot_{8n}$ being applied to the second term on the right-hand side, and then we are finally applying the predicate symbol $\|\cdot\|_{8n}$ to the overall term living in B_{8n}, or we might simply apply the metric symbol d_{8n} to these two terms.[11]

Before moving on to discuss other axioms, we should note that the previous axiom is a nightmare to read (and to write!). Thus, we will be intentionally sloppy moving forward and simply write axioms without worrying so much about including the appropriate labels for the symbols.

Returning to the list of axioms for Banach spaces, we should include an axiom connecting the norm to the metric symbol. Consequently, for each n, we need a sentence of the form

$$\sup_{x\in B_n} |d(x,0) - \|x\||\,.$$

As promised, we have stopped annotating our symbols with letters indicating their sorts, but it is still polite to explain which sorts the quantifers are ranging over. The axiom above guarantees that the norm of a vector is the same as its distance to the constant symbol for the zero vector.

One also needs axioms expressing that the functions $\iota_{m,n}$ are isometric inclusions, and we leave it to the reader to provide their precise formulations.

One final collection of axioms is necessary; namely that the sorts B_n capture precisely the vectors of norm at most n. Indeed, while the symbol for the norm on B_n has bound n as provided by the language, it is a priori possible that there is an element x living in the sort B_m for some $m > n$ and yet x has norm at most n. We want to declare that this is the case because $x = \iota_{n,m}(y)$ for some y in sort B_n. We can capture this fact using the sentence

$$\sup_{x\in B_m} \min\left(n \dot{-} \|x\|,\ \inf_{y\in B_n} \|\iota_{n,m}(y) - x\|\right).$$

If this sentence evaluates to 0 and x is a vector in the sort B_m whose norm is strictly less than n (whence the first term in the minimum is strictly positive), than for any $\epsilon > 0$, one can find a vector $y \in B_n$ for which $\|\iota_{n,m}(y) - x\| < \epsilon$. We leave it to the reader to verify that this latter statement does indeed capture the desired behavior. (Hint: we require interpretations of sorts to be *complete* metric spaces.)

It should be clear at this point that expressing familiar axioms for natural mathematical structures can be quite painful using the precise syntactic requirements of the logic we are describing. However, once we begin to discuss the computability of these structures, it is important to know that such explicit axiomatizations exist. That

[11] That these two approaches are equivalent will be justified by the next axiom scheme we discuss.

being said, for other purposes, it might be useful to know that a class of structures is elementary, even if one is uninterested in knowing the precise axioms. In that case, the following "soft" semantic test is quite useful [1, Sect. 5]:

Theorem 2.14 *Let C be a class of L-structures. Then C is an elementary class if and only if C is closed under isomorphism, ultraproducts, and ultraroots.*[12]

2.2.6 *Other Model-Theoretic Notions*

Here, we transport some important classical model-theoretic notions to the metric setting. Throughout this discussion, we fix a continuous signature L and L-structures $\mathcal{M}$ and $\mathcal{N}$. An *embedding* $i : \mathcal{M} \to \mathcal{N}$ is a function such that $i \circ f^{\mathcal{M}} = f^{\mathcal{N}} \circ i$ and $P^{\mathcal{M}} = P^{\mathcal{N}} \circ i$ for all function symbols f and predicate symbols P respectively. Note that embeddings between structures are necessarily isometric embeddings of the underlying metric spaces. A surjective embedding is called an *isomorphism*. If there is an isomorphism $i : \mathcal{M} \to \mathcal{N}$, we say that $\mathcal{M}$ and $\mathcal{N}$ are *isomorphic* and write $\mathcal{M} \cong \mathcal{N}$. We note that isomorphic structures are automatically isometric. This conflicts with the classical use of "isomorphic" for some metric structures. For example, two Banach spaces are often said to be isomorphic if they are related by a linear *homeomorphism*. However, in model theory, there is no use for such a notion of isomorphism.

$\mathcal{M}$ and $\mathcal{N}$ are said to be *elementarily equivalent*, denoted $\mathcal{M} \equiv \mathcal{N}$, if $\sigma^{\mathcal{M}} = \sigma^{\mathcal{N}}$ for all L-sentences σ. Equivalently, $\mathcal{M} \equiv \mathcal{N}$ if and only if $\mathrm{Th}(\mathcal{M}) = \mathrm{Th}(\mathcal{N})$. Note that isomorphic structures are elementarily equivalent but the converse is usually false. An important, logic-free reformulation of elementary equivalence can be given using the Keisler–Shelah theorem; namely, $\mathcal{M} \equiv \mathcal{N}$ if and only if there are ultrafilters $\mathcal{U}$ and $\mathcal{V}$ such that $\mathcal{M}^{\mathcal{U}} \cong \mathcal{N}^{\mathcal{V}}$.

An embedding $i : \mathcal{M} \to \mathcal{N}$ is called an *elementary embedding* if $\varphi^{\mathcal{M}}(\mathbf{a}) = \varphi^{\mathcal{N}}(i(\mathbf{a}))$ for all L-formulas $\varphi(\mathbf{x})$ and all tuples $\mathbf{a} \in M$. Note in particular that if there is an elementary embedding from $\mathcal{M}$ to $\mathcal{N}$, then $\mathcal{M}$ and $\mathcal{N}$ are elementarily equivalent.

If $M \subseteq N$ and the inclusion mapping yields an embedding (respectively elementary embedding) from $\mathcal{M}$ to $\mathcal{N}$, then we say that $\mathcal{M}$ is a *substructure* (respectively *elementary substructure*) of $\mathcal{N}$, denoted $\mathcal{M} \subseteq \mathcal{N}$ (respectively $\mathcal{M} \preceq \mathcal{N}$).

An important result is that "small" elementary substructures always exist. A precise statement of this result is the following.

Theorem 2.15 (Downward Löwenheim–Skolem Theorem) *Given an L-structure $\mathcal{M}$ and $X \subseteq M$, there is an elementary substructure $\mathcal{N} \preceq \mathcal{M}$ such that $X \subseteq N$ and $\mathfrak{c}(N) \leq \mathfrak{c}(X) + \mathfrak{c}(L)$. In particular, if L is a separable language, then every L-structure has a separable elementary substructure.*

Finally, we mention an important theorem relating the values of formulas in an ultraproduct and their values in the constituent structures. First, we explain how to

[12] Here, C is *closed under ultraroots* means that if $\mathcal{M}^{\mathcal{U}}$ belongs to C, then so does $\mathcal{M}$.

take the ultraproduct of a family of structures. Fix a family $(\mathcal{M}_i)_{i\in I}$ of L-structures and an ultrafilter $\mathcal{U}$ on I. Let $M = \prod_{\mathcal{U}} M_i$ denote the metric ultraproduct of the underlying metric spaces of the family.[13] We make M the underlying metric space of an L-structure $\mathcal{M}$ by defining, for n-ary function symbols f and n-ary predicate symbols P respectively,

$$f^{\mathcal{M}}(\mathbf{a}_{\mathcal{U}}) = (f^{\mathcal{M}_i}(\mathbf{a}_i))_{\mathcal{U}}$$

and

$$P^{\mathcal{M}}(\mathbf{a}_{\mathcal{U}}) = \lim_{\mathcal{U}} P^{\mathcal{M}_i}(\mathbf{a}_i).$$

We leave it to the reader to check that these definitions make sense and do in fact yield an L-structure. If each $\mathcal{M}_i$ is a fixed L-structure $\mathcal{M}$, then we refer to the ultraproduct as the *ultrapower of $\mathcal{M}$ with respect to the ultrafilter $\mathcal{U}$*, denoted $\mathcal{M}^{\mathcal{U}}$. Note that the diagonal embedding $M \hookrightarrow M^{\mathcal{U}}$ yields an embedding $\mathcal{M} \hookrightarrow \mathcal{M}^{\mathcal{U}}$.

Theorem 2.16 (Łos's Theorem) *Suppose $\mathcal{M}$ is an ultraproduct $\prod_{\mathcal{U}} M_i$ of L-structures. Then, for every L-formula $\varphi(\mathbf{x})$ and tuple $\mathbf{a} \in M$,*

$$\varphi^{\mathcal{M}}(\mathbf{a}) = \lim_{\mathcal{U}} \varphi^{\mathcal{M}_i}(\mathbf{a}_i).$$

As a consequence of Łos's Theorem, we have that the diagonal embedding is an elementary embedding.

2.3 Compactness and Completeness

A foundational result in classical logic is the Compactness Theorem, which states that a set of sentences is satisfiable if and only if it is finitely satisfiable. The same result holds in continuous logic. In fact, satisfiability is equivalent to the a priori weaker notion of approximate finite satisfiability:

Theorem 2.17 (Compactness Theorem) *Suppose that Γ is an L-theory. Then the following are equivalent.*

1. *Γ is satisfiable.*
2. *Γ is* finitely satisfiable, *that is, every finite subset of Γ is satisfiable.*
3. *Γ is* approximately finitely satisfiable, *that is, every finite subset of*

$$\left\{\sigma \dot{-} \frac{1}{n} \;:\; \sigma \in \Gamma, n \geq 1\right\}$$

is satisfiable.

[13] Technically, we only defined this for families indexed by $\mathbb{N}$, but the general definition is entirely the same.

The importance of the equivalence between satisfiability and approximate finite satisfiability is that to show that Γ has a model, one only needs to show that, given any finite subset F of Γ and $\epsilon > 0$, there is an L-structure $\mathcal{M}$ such that $\sigma^{\mathcal{M}} < \epsilon$ for all $\sigma \in F$.

As with classical logic, one can prove the Compactness Theorem using ultraproducts. The idea for the nontrivial direction (3) implies (1) is as follows. Let I be the set of pairs (F, ϵ) where $F \subseteq \Gamma$ is finite and $\epsilon > 0$. For each $\sigma \in \Gamma$ and $\epsilon > 0$, let $X_{\sigma,\epsilon} = \{(F, \epsilon') \in I \ : \ \sigma \in F, \epsilon' \leq \epsilon\}$. It is easy to show that the collection of $X_{\sigma,\epsilon}$'s has the finite intersection property, whence there is an ultrafilter $\mathcal{U}$ on I for which each $X_{\sigma,\epsilon}$ belongs to $\mathcal{U}$. Since Γ is approximately finitely satisfiable, for each $(F, \epsilon) \in I$, there is an L-structure $\mathcal{M}_{(F,\epsilon)}$ such that $\sigma^{\mathcal{M}_{(F,\epsilon)}} < \epsilon$ for all $\sigma \in F$. Set $\mathcal{M} = \prod_{\mathcal{U}} \mathcal{M}_{(F,\epsilon)}$. Since $(F, \epsilon) \in X_{\sigma,\epsilon}$ for each $\sigma \in F$, it follows from Łos' Theorem that $\sigma^{\mathcal{M}} \leq \epsilon$ for all $\sigma \in \Gamma$ and $\epsilon > 0$. Since $\epsilon > 0$ is arbitrary, we have that $\mathcal{M}$ is the desired model of Γ.

Another foundational result in first-order logic is Gödel's Completeness Theorem, which asserts that logical implication $\Gamma \models \sigma$ implies deducibility via formal proof $\Gamma \vdash \sigma$.[14] There is an analogous result in continuous logic. Indeed, Ben Yaacov and Pederson [4] developed a proof system for continuous logic in which it makes sense to say that a continuous sentence σ is derivable from a set Γ of continuous sentences, denoted $\Gamma \vdash \sigma$. As in classical logic, an important observation is that if Γ is a computably enumerable (c.e.) set of L-sentences (in some computable language L[15]), then so is the set $\{\sigma \ : \ \Gamma \vdash \sigma\}$ of consequences of Γ. With a proof system at hand, one can then prove the usual Soundness and Completeness Theorems: Γ is *consistent*, that is, Γ does not prove every L-sentence, if and only if Γ is satisfiable. An important corollary of their work is the following.

Theorem 2.18 (Pavelka-style completeness) *For any L-theory Γ and L-sentence σ, we have*

$$\sup\{\sigma^{\mathcal{M}} \ : \ \mathcal{M} \models \sigma\} = \inf\{r \in \mathbb{Q}^{>0} \ : \ \Gamma \vdash \sigma \dot{-} r\}.$$

The inequality $\leq$ is a sort of soundness theorem: if $\Gamma \vdash \sigma \dot{-} r$, then Γ proves that σ is no more than r, whence $\sigma^{\mathcal{M}} \leq r$ for all models $\mathcal{M}$ of σ. The inequality $\geq$ is really an approximate completeness assertion: if $r \in \mathbb{R}^{>0}$ and $\sigma^{\mathcal{M}} \leq r$ for all $\mathcal{M} \models \Gamma$, then there must be a "good reason" for this; that is, Γ can prove that σ is at most q for every $q \in \mathbb{Q}^{>0}$ with $r \leq q$. Pavelka-style completeness has been a key component in the proofs of many computability results in operator algebras (see Chaps. 7 and 8).

2.4 Some Nuances

The general impression we have given in the previous sections is that continuous logic is a rather straightforward generalization of classical logic and that, after some practice, it becomes rather routine to generalize many important notions and theorems

14 The easier to prove Soundness Theorem provides the converse statement.

15 This will be made more precise in the next chapter.

from classical logic to the continuous context. In this section, we want to mention a few nuances which highlight phenomena that appear in continuous logic that have no classical counterpart.

2.4.1 General Formulas

In this subsection, we fix a signature L and an L-theory T.

There is a natural pseudometric ρ_T on L-formulas in the free variables $\mathbf{x}$ given by

$$\rho_T(\varphi(\mathbf{x}), \psi(\mathbf{x})) = \sup\left\{|\varphi^{\mathcal{M}}(\mathbf{a}) - \psi^{\mathcal{M}}(\mathbf{a})| \ : \ \mathcal{M} \models T, \ \mathbf{a} \in M\right\}.$$

Note then that φ and ψ are *T-equivalent*, that is, $\varphi^{\mathcal{M}}(\mathbf{a}) = \psi^{\mathcal{M}}(\mathbf{a})$ for all models $\mathcal{M}$ of T and all tuples $\mathbf{a}$ from M, if and only if $\rho_T(\varphi, \psi) = 0$. Consequently, we should identify T-equivalent formulas. For the sake of the kinds of constructions we encounter in analysis, we form the completion of this space.

Definition 2.19 A *T-formula* in the variables $\mathbf{x}$ is an element of the ρ_T-completion of the pseudometric space of formulas in the variables $\mathbf{x}$.

For example, given any countable collection $\varphi_n(\mathbf{x})$ of L-formulas, the infinite sum $\sum_n \frac{\varphi_n}{2^n R_\varphi}$ is a T-formula for any L-theory T. (Recall that R_φ is the bound on the formula φ known to the signature.)

The key observation is that one can interpret T-formulas in models of T. Namely, if φ is a T-formula represented by the ρ_T-Cauchy sequence $(\varphi_n)_{n\in\mathbb{N}}$ and $\mathcal{M} \models T$, one can set $\varphi^{\mathcal{M}} = \lim_n \varphi_n^{\mathcal{M}}$. The reader can verify that this definition is independent of the choice of representative sequence. Moreover, each $\varphi^{\mathcal{M}}$ is uniformly continuous and, as usual, in a way known to T; that is, there is a single modulus of uniform continuity that works for each $\varphi^{\mathcal{M}}$ as $\mathcal{M}$ ranges over models of T.

2.4.2 Definable Sets

Given the central importance of the notion of definable set in classical model theory, it is natural to ask for the correct notion of definable set in continuous logic. The material from this section was taken from Goldbring's article [5].

To be clear, in this context, we are speaking of global definability with respect to a theory T (analogous to how a formula $\varphi(\mathbf{x})$ in classical logic can be viewed as an assignment $\mathcal{M} \mapsto \varphi(\mathcal{M})$ from models of T to subsets of cartesian powers of models of T). There is also a local version of definability relative to a structure that we will not discuss here.

Given that definable sets in classical logic are just given by looking at the set of elements where a given formula is true, the naïve guess might be as follows: given a T-formula φ, the set it defines should just be the zeroset of φ given by

$$Z(\varphi)(\mathcal{M}) = \{a \in M \ : \ \varphi^{\mathcal{M}}(a) = 0\}.$$

By using our earlier tricks, we can see that this suggestion is equivalent to asking that the various level sets of φ be definable sets. Unfortunately, this notion of definability is not robust enough to yield a workable theory. In particular, quantification of a T-formula over a definable set does not lead to a T-formula in general.

The key to isolating the correct notion of definable set is the following theorem. Below, by a *T-functor X over* $\mathbf{x}$, we simply mean a functor $\mathcal{M} \mapsto X(\mathcal{M})$ from models $\mathcal{M}$ of T to closed subsets of $M^{\mathbf{x}}$ which behaves appropriately under substructure; namely, if $\mathcal{M} \subseteq \mathcal{N}$, then $X(\mathcal{M}) = X(\mathcal{N}) \cap M^{\mathbf{x}}$, and X maps the inclusion map to the corresponding inclusion map. The zerosets above are examples of T-functors. Also, a *T-function Φ over* $\mathbf{x}$ is simply an assignment to each model $\mathcal{M}$ of T a function $\Phi^{\mathcal{M}} : M^{\mathbf{x}} \to \mathbb{R}$. Examples of T-functions are given by the interpretation of T-formulas, and we call such T-functions *realized*.

Theorem 2.20 *Suppose that X is a T-functor over* $\mathbf{x}$. *Then the following are equivalent.*

1. *For all T-formulas $\psi(\mathbf{x}, \mathbf{y})$, the T-functions* $\sup_{\mathbf{x}\in X} \psi(\mathbf{x}, \mathbf{y})$ *and* $\inf_{\mathbf{x}\in X} \psi(\mathbf{x}, \mathbf{y})$ *are realized by T-formulas.*
2. *The T-function $d(\mathbf{x}, X)$ is realized by a T-formula.*
3. *For all sets I, all families of models $(\mathcal{M}_i)_{i\in I}$ of T, and all ultrafilters $\mathcal{U}$ on I, we have $X(\prod_{\mathcal{U}} \mathcal{M}_i) = \prod_{\mathcal{U}} X(\mathcal{M}_i)$.*
4. *There is an* almost-near T-formula $\varphi(\mathbf{x})$ *such that* $X = Z(\varphi)$.

In item (4) above, by an almost-near T-formula φ we mean a T-formula with the property that for all $\epsilon > 0$, there is $\delta > 0$ such that, for all models $\mathcal{M}$ of T and $\mathbf{a} \in M$, if $\varphi^{\mathcal{M}}(\mathbf{a}) < \delta$, then there is $\mathbf{b} \in M$ such that $d(\mathbf{a}, \mathbf{b}) \leq \epsilon$ and $\varphi^{\mathcal{M}}(\mathbf{b}) = 0$.

Hopefully it is evident to the reader that the properties listed in items (1)–(3) are all desirable properties of a T-functor. Item (1) expresses the fact that T-formulas are closed under quantification over T-functors, (2) expresses that the distance to a T-functor is given by a T-formula, and (3) expresses that T-functors commute with ultraproducts. These properties are equivalent to the property of being a zeroset of an almost-near formula, so the naïve guess of being a zeroset was not too far off the mark! The T-functors satisfying the equivalent properties of the previous theorem are called *T-definable sets*.

Remark 2.21 A particular consequence of item (1) says that if $\psi(\mathbf{x}, \mathbf{y})$ is an L-formula and $\varphi(\mathbf{x})$ is an almost-near T-formula, then for every $\epsilon > 0$, there is an L-formula $\theta(\mathbf{y})$ such that $|\sup_{\mathbf{x}\in Z(\varphi)} \psi(\mathbf{x}, \mathbf{y}) - \theta(\mathbf{y})| < \epsilon$. It can be shown that if φ has a computable modulus of uniform continuity, then the assignment $\epsilon \mapsto \theta$ above can be made effectively, a fact that is used in Chap. 8 below.

As briefly mentioned above, there is a notion of a subset of a particular structure being definable; one can describe this notion in terms of the global notion just considered, allowing one to prove a version of Theorem 2.20 in this local context. We refer the reader to [5] for more on this. Note that definability relative to a structure is what will be needed for our applications in Sect. 8.36 below.

2.4.3 *Type Spaces*

We now assume that our L-theory T is complete. The notion of a complete type is fairly easy to transport to the continuous setting.

Definition 2.22 Given $\mathcal{M} \models T$ and $\mathbf{a} \in M$, the *complete type of* $\mathbf{a}$ *in* $\mathcal{M}$ is the function $\mathrm{tp}^{\mathcal{M}}(\mathbf{a})$ defined on the set of T-formulas by $\mathrm{tp}^{\mathcal{M}}(\mathbf{a})(\varphi) = \varphi^{\mathcal{M}}(\mathbf{a})$. The set of complete types is denoted $S_{\mathbf{x}}(T)$.

Remark 2.23 Since the T-formulas given by L-formulas are dense in the set of all T-formulas, a complete type is determined by its restriction to the set of L-formulas. Also, the complete type of a tuple is determined by its zeroset.

If $p \in S_{\mathbf{x}}(T)$, $\mathcal{M} \models T$, and $\mathbf{a} \in M^{\mathbf{x}}$, we say that $\mathbf{a}$ *realizes* p, denoted $\mathbf{a} \models p$, if $p = \mathrm{tp}^{\mathcal{M}}(\mathbf{a})$. We also let $p(\mathcal{M})$ denote the set of realizations of p in $\mathcal{M}$. We note that the assignment $\mathcal{M} \mapsto p(\mathcal{M})$ is a T-functor.

There is a natural topology on $S_{\mathbf{x}}(T)$, called the *logic topology*, which has subbasic open sets of the form $\{p \in S_{\mathbf{x}}(T) \ : \ p(\varphi) \in (a, b)\}$ as φ ranges over L-formulas and (a, b) ranges over open intervals of $\mathbb{R}$. As in classical logic, using the Compactness Theorem, one sees that $S_{\mathbf{x}}(T)$ is a compact Hausdorff space. However, unlike in classical logic, $S_{\mathbf{x}}(T)$ is no longer totally disconnected. By the Stone–Weierstrass Theorem, the continuous functions from $S_{\mathbf{x}}(T)$ into $\mathbb{R}$ are precisely those of the form $p \mapsto p(\varphi)$ for some T-formula φ (yet another reason for considering T-formulas).

A completely new phenomenon appears in continuous logic as the type spaces now carry a second topology stemming from a natural metric.

Definition 2.24 If T is a complete theory, the metric d on $S_{\mathbf{x}}(T)$ is defined by setting

$$D(p, q) = \inf \{d(a, b) \ : \ \mathcal{M} \models T, \ a, b \in M, \ a \models p, b \models q\}.$$

Note that the completeness of the theory ensures that $D(p, q)$ is finite for all pairs (p, q) from $S_{\mathbf{x}}(T)$. While it is easy to see that D is symmetric and reflexive, it takes quite a bit of work to show that d satisfies the triangle inequality.

Uniform continuity of formulas shows that every logically open set is also metrically open. It is natural to inquire when these two topologies are actually the same. The statement that these two topologies coincide has significant model-theoretic content.

Theorem 2.25 (Ryll–Nardzewski Theorem for continuous logic) *For a complete theory T in a separable language, the following are equivalent.*

1. *The logic and metric topologies on $S_{\mathbf{x}}(T)$ agree for all* $\mathbf{x}$.
2. *The metric topology on $S_{\mathbf{x}}(T)$ is compact for all* $\mathbf{x}$.
3. *T is* $\aleph_0$-categorical, *that is, T has a unique separable model.*

Rather than asking for the two topologies to agree globally, one can also consider the question of when the two topologies agree at a particular type. The statement that the logic and metric topologies agree at a given type also carries significant model-theoretic content. But first, we need a definition.

Definition 2.26 $p \in S_{\mathbf{x}}(T)$ is *principal* if the T-functor p is a T-definable set.

Theorem 2.27 *The following are equivalent.*

1. *p is principal.*
2. *The logic and metric topologies agree at p.*
3. *Every metric ball around p has nonempty interior in the logic topology.*

One can show that principal types are realized in all models $\mathcal{M}$ of T. The converse holds as well.

Theorem 2.28 (Omitting Types Theorem) *If T is a complete theory in a separable language, and if p is a nonprincipal type, then there is a separable model $\mathcal{M}$ of T* omitting *p, that is, $p \neq \mathrm{tp}^{\mathcal{M}}(\mathbf{a})$ for any tuple* $\mathbf{a}$ *from M.*

References

1. Itaï Ben Yaacov, Alexander Berenstein, C. Ward Henson, and Alexander Usvyatsov. Model theory for metric structures. *London Mathematical Society Lecture Note Series*, 350:315, 2008.
2. Bradd Hart. An introduction to continuous model theory. *Model Theory of Operator Algebras*, 11:83, 2023.
3. Isaac Goldbring and Bradd Hart. The universal theory of the hyperfinite II_1 factor is not computable. *Bull. Symb. Log.*, 30(2):181–198, 2024.
4. Itaï Ben Yaacov and Arthur Paul Pedersen. A proof of completeness for continuous first-order logic. *The Journal of Symbolic Logic*, 75(1):168–190, 2010.
5. Isaac Goldbring. Spectral gap and definability. In *Beyond First Order Model Theory, Volume II*, pages 103–138. Chapman and Hall/CRC, 2023.

Chapter 3
Framework for the Computability of Metric Structures

Having set forth the model theory for metric structures, we now lay out a framework for the computability of these structures based on their model theory. This framework is similar to the approach in [1] and is a generalization of the approaches to Banach spaces in [17]. It also generalizes the approach to C*-algebras set forth in [2].

3.1 A Short Survey of Classical Computable Structure Theory

For the sake of motivating our work in this chapter, we define the main themes and state some of the foundational results of classical computable structure theory. For a more expansive introduction, we refer the reader to [3, 4].

There are two main approaches to computable structure theory: one western and one eastern. The western approach studies structures whose domain is the natural numbers and whose relations and operations are uniformly computable. The eastern approach considers *presentations* of structures which consist of labeling elements of a structure with natural numbers. The approaches are equivalent. However, as the eastern approach seems to extend more naturally to the study of metric structures, it is the approach we will follow here.

3.1.1 Computable Presentations

We begin with an introduction to presentations and a discussion of accompanying terminology.

J. N. Y. Franklin et al., *Effective Metric Structure Theory*, Springer Monographs in Mathematics, https://doi.org/10.1007/978-3-032-13322-9_3

Definition 3.1 Suppose $\mathcal{A}$ is a (classical) structure.

1. A *notation* for $\mathcal{A}$ is a surjection of $\mathbb{N}$ onto A.
2. If ν is a notation for $\mathcal{A}$, then the pair $(\mathcal{A}, \nu)$ is called a *presentation* of $\mathcal{A}$. It is also referred to as a *presented structure*.

We use expressions such as $\mathcal{A}^\#$ and $\mathcal{A}^\dagger$ for presentations of a structure $\mathcal{A}$.

Definition 3.2 If $\mathcal{A}^\# = (\mathcal{A}, \nu)$ is a presented structure, then n shall be referred to as an *$\mathcal{A}^\#$-notation* for $\nu(n) \in A$.

We now assume that the language L of each structure $\mathcal{A}$ considered has an associated Gödel numbering. This is a surjection γ of the natural numbers onto L that satisfies the following conditions:

- From n, it is possible to compute the type of $\gamma(n)$, that is, it can be effectively determined if $\gamma(n)$ is a constant, function, or relation symbol.
- From n, it is possible to compute the arity of $\gamma(n)$.

We call n a *code* of $\gamma(n)$. Languages that admit such a coding are called *computable*.

Now we can discuss the effectivization of these concepts.

Definition 3.3 Suppose $\mathcal{A}^\#$ and $\mathcal{B}^\#$ are presented structures and $f : A^n \to B$. We say that f is a *computable map from $\mathcal{A}^\#$ to $\mathcal{B}^\#$* if there is an algorithm that, given any $\mathcal{A}^\#$-notations for $a_1, \ldots, a_n \in A$, computes a $\mathcal{B}^\#$-notation for $f(a_1, \ldots, a_n)$.

Definition 3.4 Suppose $\mathcal{A}^\#$ is a presented structure and let R be an n-ary relation on A. We say that R is a *computable relation of $\mathcal{A}^\#$* if there is an algorithm that, given any $\mathcal{A}^\#$-notations for $a_1, \ldots, a_n \in A$, determines if $R(a_1, \ldots, a_n)$ holds.

Definition 3.5 Suppose $\mathcal{A}^\#$ is a presented structure. We say $\mathcal{A}^\#$ is *computable* if it satisfies the following conditions:

1. There is an algorithm that, given any two natural numbers, determines if they are $\mathcal{A}^\#$-notations for the same element of A.
2. Each function f of $\mathcal{A}$ is a computable function of $\mathcal{A}^\#$ uniformly in (a code of) f.
3. Each relation R of $\mathcal{A}$ is a computable relation of $\mathcal{A}^\#$ uniformly in (a code of) R.

If $\mathcal{A}^\#$ is computable, then an *index* of $\mathcal{A}^\#$ is any natural number that codes Turing machines that effect the algorithms in Definition 3.5. We note that there is no need to include a condition for the computability of constants, as they are represented by natural numbers.

It should be noted that not everyone includes condition (1) in the definition of a computable structure. Presentations for which this relation is c.e. are also studied. However, if this relation is co-c.e., then there is a presentation for which it is computable.

As said above, the approach presented here is the eastern approach. In the western approach, by a *computable structure* one means a structure whose domain is the natural numbers and whose functions and relations are uniformly computable. As every finite structure has a computable presentation, a computable presentation of a structure $\mathcal{A}$ is often defined as an isomorphism of a computable structure with $\mathcal{A}$. Definition 3.5, however, is closer in spirit to the definition we shall use for the computable presentation of a metric structure.

Definition 3.6 A structure is *computably presentable* if it has at least one computable presentation.

3.1.2 Computable Categoricity and Related Notions

Fröhlich and Shepherdson constructed a field with two computable presentations so that one has a splitting algorithm and the other does not [5]. Thus, how one denotes the elements of a structure can affect what can be computed on the structure. This observation motivates the definition of computable categoricity.

Definition 3.7 Let $X \subseteq \mathbb{N}$.

1. A structure is X*-computably categorical* if any two of its computable presentations are X-computably isomorphic, that is, X can compute an isomorphism between any two of its computable presentations.
2. A *computably categorical* structure is a $\emptyset$-computably categorical structure.
3. If $X = \emptyset^{(n-1)}$, we say that the structure is Δ^0_n*-categorical.*

Example 3.8 $(\mathbb{Q}, <)$ is computably categorical. This can be seen by effectivizing Cantor's back-and-forth technique.

Example 3.9 $(\mathbb{N}, <)$ is not computably categorical. In fact, it is fairly easy to construct a presentation of this order in which the successor relation is not computable.

The property of computable categoricity was referred to as *autostability* by Maltsev. This terminology is still used by some authors. However, the terminology of computable categoricity is, at the time of this writing, more frequently used.

There are at least two ways to gauge how far a structure is from being computably categorical: its degree of categoricity (if defined) and its computable dimension.

Definition 3.10 The *degree of categoricity* of a structure A is the minimal Turing degree (if it exists) that computes an isomorphism between any two computable presentations of A.

We can thus say that the degree of categoricity of a structure calibrates the structure's complexity. Degrees of categoricity were first defined by Fokina et al. [6]. They showed that if $\mathbf{0}^{(m)} \leq_T \mathbf{d}$ and $\mathbf{d}$ is c.e. or d-c.e. in $\mathbf{0}^{(m)}$, then $\mathbf{d}$ is a degree of categoricity. Csima, Franklin, and Shore later extended this result through the hyperarithmetic hierarchy and showed that every degree of categoricity is hyperarithmetic [7]. Fokina, Kalimullin, and R. Miller also considered certain types of structures commonly investigated in classical computable structure theory. For instance, they proved that every c.e. degree is the degree of categoricity of some computably presentable algebraic field [6].[1]

Definition 3.11 The *computable dimension* of a structure is the number of its computable presentations up to computable isomorphism.

The following result of Goncharov connects these two gauges (see [3] for a proof).

Theorem 3.12 *If a structure $\mathcal{A}$ is $\mathbf{0}'$-categorical, then its computable dimension is* 1 *or* $\aleph_0$.

Khoussainov, Hirschfeldt, Shore, and Slinko have constructed examples of structures with intermediate computable dimension within a wide range of natural classes of algebraic structures [8].

3.1.3 Index Sets

Classification problems are ubiquitous in mathematics. One method of classifying structures from some class is to associate each of them with an invariant from some well-understood class of structures such as the natural numbers or a countable group. Another is to provide a method for determining whether two finitely defined structures are isomorphic. Mathematical logic offers two approaches for assessing the difficulty of a classification problem: descriptive set theory and computability theory. Descriptive set theory seeks to place the isomorphism problem for a class of structures in the Borel hierarchy. Computability theory uses indices of presentations, which are regarded as finite descriptions of structures, to place the isomorphism problem in the hyperarithmetical hierarchy. The latter considerations lead to the following definitions.

Definition 3.13 Let $\mathcal{K}$ be a class of structures (for example, the class of all groups).

1. The *index set of* $\mathcal{K}$ is the set of all natural numbers that index presentations of elements of $\mathcal{K}$.
2. The *isomorphism problem of* $\mathcal{K}$ is the set of all pairs $(e, e') \in \mathbb{N}^2$ so that e and e' index presentations of isomorphic structures in $\mathcal{K}$.
3. The *embedding problem of* $\mathcal{K}$ is the set of all pairs $(e, e') \in \mathbb{N}^2$ for which there exist presentations $\mathcal{A}^\#$ and $\mathcal{B}^\dagger$ of structures in $\mathcal{K}$ so that e indexes $\mathcal{A}^\#$, e' indexes $\mathcal{B}^\dagger$, and $\mathcal{A}$ embeds into $\mathcal{B}$.

For example, it is shown in [9] that the isomorphism problem for vector spaces over $\mathbb{Q}$ is Π^0_3-complete. This makes intuitive sense since there is a fairly simple

[1] A field is *algebraic* if it is a subfield of the algebraic closure of $\mathbb{Q}$ or of the algebraic closure of $\mathbb{Z}_p$ for some prime p.

classification of these spaces. At the same time, in the same paper, it is shown that the isomorphism problem for ordered real closed fields is Σ^1_1-complete. This implies that there is *no* arithmetical classification theorem for ordered real closed fields. More examples can be found in [10].

3.2 Presentations of Metric Structures

In the last chapter, we discussed how classical model theory is adapted to the setting of metric structures, while in the previous section we briefly surveyed the classical setting of computable structure theory. We now begin to discuss how the tools of computable analysis are used to extend classical computable structure theory to the realm of (usually uncountable) metric structures.

To understand the key obstacle in this extension, let us first recall that classical computable structure theory begins with the concept of a computable presentation of a structure in which the elements of the structure are labeled by natural numbers in such a way that the pullback functions and relations are computable. In particular, since no number labels more than one structural element, this approach cannot directly be applied to uncountable structures.

At the same time, most scientific computation involves continuous data. Fortunately, computable analysis provides a sound scientific foundation for the study of computation on uncountable, but separable, metric structures. Namely, with respect to some well-chosen dense set, one emphasizes effective *approximation* of objects rather than their exact computation in a finite number of steps. This is the approach we will follow here.

Fix a countable continuous language L, which is possibly many sorted.[2] Assume further that the sorts are enumerated as $S_0, S_1, \ldots, S_n$ (if there are finitely many sorts) or $S_0, S_1, S_2, \ldots$ (if there are infinitely many sorts). We say that L is *computable* if there is a surjection $\nu : \mathbb{N} \to L$ that satisfies the following criteria:

- From n, it is possible to compute the type of $\nu(n)$, that is, it can be effectively determined if $\nu(n)$ is a constant, function, or predicate symbol.
- From n, it is possible to determine the domain and codomain of $\nu(n)$.
- From n it is possible to determine the modulus of uniform continuity for $\nu(n)$; that is, given $n, k \in \mathbb{N}$, it is possible to compute $m \in \mathbb{N}$ so that $\Delta_{\nu(n)}(2^{-k}) \geq 2^{-m}$.

As in the classical case, we call ν a *Gödel numbering (or labeling)* of L and refer to n as a *code* of $\nu(n)$.

It is easy to see that each particular continuous language considered in the previous chapter is computable in this sense. Of course, some of the languages previously considered are uncountable. As pointed out in Remark 2.6, in these cases, we can work with countable "dense" sublanguages. For example, when representing a Banach

[2] We did not consider many-sorted languages in our brief discussion of classical computable structure theory in the previous section, but doing so would have added no further complications.

space as a metric structure, it is sufficient to have one scalar multiplication function for each rational scalar.

Note that a Gödel numbering of a computable language L can be "extended" to a Gödel numbering of the L-terms in such a way that:

- From codes for a function symbol f and terms $t_1, \ldots, t_n$ of the appropriate kind, one can determine a code for $f(t_1, \ldots, t_n)$.
- One can effectively determine a modulus of uniform continuity for each term t.

Throughout this chapter, we assume that L is a computable continuous language, and we fix an L-structure $\mathcal{M}$. We now work towards defining what we mean by a presentation of $\mathcal{M}$. We require a few preliminaries.

To begin, we assume that our L-structures arise as dissections of meta-structures (see Example 2.5 for a discussion of this concept). Thus, we assume that $S_j^{\mathcal{M}} \subseteq S_{j+1}^{\mathcal{M}}$ and $M = \bigcup_j S_j^{\mathcal{M}}$. In addition, we assume that the sorts of $\mathcal{M}$ are *strongly increasing* in that $S_j^{\mathcal{M}}$ is included in the interior of $S_{j+1}^{\mathcal{M}}$. To illustrate what we have in mind, we return to one of the examples of Chapter 2. Suppose $\mathcal{B}$ is a Banach algebra. When we represent $\mathcal{B}$ in the framework of Chapter 2, we may define the jth sort to be $B(\mathbf{0}; 2^j)$. The modulus of continuity for addition is the same for each type. However, the modulus of continuity for multiplication on the jth sort is given by $k \mapsto k + j$.

We emphasize that these assumptions are for ease of exposition only. We also emphasize that these assumptions are satisfied by all the examples we considered in Chapter 2 and all the structures we consider in subsequent chapters. In Sect. 3.8, we outline how to modify our framework without these restrictions. Only the proof of Theorem 3.50 below significantly uses these assumptions, and in Sect. 3.8 we will discuss how to rephrase this result so as to accommodate their absence. We note there are some similarities between what we are doing here and the domains of quantification approach of Lupini [11].

Now, for $S \subseteq M$, let $\langle S \rangle$ denote the *presubstructure* generated by S, that is, the smallest subset of M that contains S and is closed under the function symbols of L. The closure of $\langle S \rangle$, denoted $\overline{\langle S \rangle}$, is then the substructure of $\mathcal{M}$ generated by S.

Definition 3.14 Let $\mathcal{M}$ be a metric structure, and let $(p_n)_{n\in\mathbb{N}}$ be a sequence of points of $\mathcal{M}$. We say $(p_n)_{n\in\mathbb{N}}$ is a *generating sequence* for $\mathcal{M}$ if $\overline{\langle\{p_n \ : \ n \in \mathbb{N}\}\rangle} = M$.

Definition 3.15

1. A *presented metric structure* is a pair $(\mathcal{M}, (p_n)_{n\in\mathbb{N}})$ where $\mathcal{M}$ is a metric structure and $(p_n)_{n\in\mathbb{N}}$ is a generating sequence for $\mathcal{M}$.
2. If $(p_n)_{n\in\mathbb{N}}$ is a generating sequence for a metric structure $\mathcal{M}$, then $(\mathcal{M}, (p_n)_{n\in\mathbb{N}})$ is called a *presentation of* $\mathcal{M}$.

Just as with classical structures, we use notations such as $\mathcal{M}^\#$ and $\mathcal{M}^\dagger$ to denote presentations of a metric structure $\mathcal{M}$.

We regard a presented metric structure as a kind of metric structure. Thus, if $\mathcal{M}^\#$ is a presentation of $\mathcal{M}$, then $\mathcal{M}^\#$ inherits all of the objects associated with $\mathcal{M}$. For example, each functional[3] of $\mathcal{M}$ will be referred to as a functional of $\mathcal{M}^\#$ and each operation of $\mathcal{M}$ will be referred to as an operation of $\mathcal{M}^\#$.

Definition 3.16 Suppose $\mathcal{M}^\# = (\mathcal{M}, (p_n)_{n\in\mathbb{N}})$ is a presented metric structure.

1. We call p_n the nth *distinguished point of* $\mathcal{M}^\#$.
2. Each point in $\langle\{p_n \ : \ n \in \mathbb{N}\}\rangle$ is called a *generated point of* $\mathcal{M}^\#$.

The Gödel numbering of the L-terms gives rise to a Gödel numbering of the generated points of $\mathcal{M}^\#$ with the following property: from a code of a function symbol f of L and codes of generated points $p_1, \dots, p_n$ of $\mathcal{M}^\#$ of the appropriate kind, it is possible to compute a code of $f^{\mathcal{M}}(p_1, \dots, p_n)$. To simplify our exposition, we identify each generated point with each of its codes.

Suppose that $\mathcal{M}$ is a (one-sorted, for simplicity) classical structure viewed as a metric structure by equipping it with the discrete metric. In this case, a presentation of $\mathcal{M}$ as a continuous structure is really a labeling not of M as in the classical sense but rather of a generating subset of M. The generated points of this (continuous) presentation then yield a presentation of $\mathcal{M}$ in the classical sense of the word.

Of course, one could have emulated classical computable structure theory by defining a presentation to be a labeling of a generating set for the structure and noting that the entire structure can then be labeled by using labels for terms in the generators, just as we do here. Then the theory we are proposing here would be a perfect generalization of its classical counterpart. Alternatively, one could do away with the difference between distinguished and generated points in the theory we are currently developing by demanding that the distinguished points be closed under the function symbols and dense in the structure. The theory we are about to develop would be unaffected by this demand, as a presented structure in the current sense would be "computably isomorphic" (in a manner to be defined later in this chapter) to the presentation obtained by replacing the distinguished points by the generated points. Nevertheless, we propose to continue with the above definition, as it allows us extra flexibility when defining and working with presentations. In Chapter 7, it will be particularly useful when dealing with universal C*-algebras.

With certain metric structures, there is associated a presentation that is designated as *standard*. In such a case, we identify the structure and its standard presentation. We now define the standard presentations of many of the metric structures commonly encountered in classical mathematics. Each of these structures is a normed space, and we view them as many-sorted structures where the nth sort is the closed ball $\overline{B}(\mathbf{0}; n+1)$.

[3] Here we use "functional" and "predicate" interchangeably. If a functional is not real valued (such as the inner product of a complex Hilbert space), then we regard it as having been resolved into its real and imaginary components.

Example 3.17 Fix a positive integer n, and let $e_0, \ldots, e_{n-1}$ be the vectors in the standard basis for $\mathbb{F}^n$. The standard presentation of $\mathbb{F}^n$ (with the Euclidean metric) is defined by declaring the mth distinguished point to be e_k where $k \in \{0, \ldots, n-1\}$ is equivalent to m modulo n. The generated points of this presentation are precisely the vectors whose components are all rational scalars.

As we shall see in Chapter 5 below, one of the Banach spaces that has received much attention in the development of EMST is the Banach space $C([0, 1]; \mathbb{F})$. The vectors of this space are the continuous functions from $[0, 1]$ into $\mathbb{F}$. The vector space operations are defined pointwise, and the norm is the supremum norm. Another space of significance for EMST is the C*-algebra $C([0, 1]; \mathbb{F})$.[4] This algebra is obtained by augmenting $C([0, 1]; \mathbb{F})$ with pointwise multiplication and the involution that is conjugation.

Example 3.18 The standard presentation of the Banach space $C([0, 1]; \mathbb{F})$ is defined by declaring the nth distinguished point to be the map $t \mapsto t^n$. This is also the standard presentation of $C^*([0, 1]; \mathbb{F})$ as a C*-algebra. The generated points of either presentation are precisely the polynomials with rational coefficients.

A *Lebesgue space* is a Banach space that is an L^p space for some p. The computability-theoretic aspects of these spaces is presented in Chapter 6. We define some of their standard presentations here.

Example 3.19 Suppose $1 \leq p < \infty$. Let $(I_n)_{n \in \mathbb{N}}$ be an effective enumeration of the dyadic subintervals of $[0, 1]$; here, by "effective" we mean that there is an algorithm that computes the endpoints of I_n from n. The standard presentation of $L^p[0, 1]$ is defined by declaring the nth distinguished point to be the characteristic function of I_n. The generated points of this presentation are precisely the rational square wavelets.

Example 3.20 Suppose $1 \leq p < \infty$. We define the standard presentation of ℓ^p by declaring the nth distinguished point to be the nth standard basis vector e_n. Thus, a vector in ℓ^p is a generated point of this presentation if and only if its support is finite and its components are all rational scalars.

Example 3.21 Let n be a positive integer and suppose $1 \leq p \leq \infty$. We define the standard presentation of ℓ_n^p by declaring the kth distinguished point to be the jth standard basis vector where $j \equiv k \mod n$. This presentation has the same distinguished and generated points as the standard presentation of $\mathbb{F}^n$. However, the norms are, of course, different if $p \neq 2$.

Up until now, we have only considered the standard presentations of classical metric structures. However, nonstandard presentations of these structures exist. The following example of a nonstandard presentation is essentially due to Pour-El and Richards [17].

---

[4] We use the same notation for considering $C([0, 1]; \mathbb{F})$ as a Banach space and as a field; we hope that context will prevent any confusion.

Example 3.22 Suppose $\mathbb{F} = \mathbb{C}$, and let $p \geq 1$ be a computable real number, that is, a real number with a computable decimal expansion. Fix a unimodular and incomputable $\zeta \in \mathbb{C}$ (that is, we assume that either the real or the imaginary part of ζ is incomputable). Define a presentation $(\ell^p)^\#$ by defining the nth distinguished point to be ζe_n. Later, we will show there are many significant differences between this presentation and the standard presentation of ℓ^p, all of which arise from the incomputability of ζ.

We now discuss products of presentations. For each $n \in \mathbb{N}$, fix a computable bijection $(\cdot)^n$ of $\mathbb{N}$ onto $\mathbb{N}^n$.

Definition 3.23 For each $j \in \{1, \ldots, n\}$, suppose $\mathcal{M}_j^\#$ is a presentation of a metric structure $\mathcal{M}_j$. Then $\mathcal{M}_1^\# \times \ldots \times \mathcal{M}_n^\#$ is the presentation of $\mathcal{M}_1 \times \ldots \times \mathcal{M}_n$ whose m^{th} distinguished point is $(p_1, \ldots, p_n)$ where p_j is the $(m)^{n^{\text{th}}}_j$ generated point of $\mathcal{M}_j^\#$.

Note that a metric structure has a presentation if and only if it is separable, in which case it actually has uncountably many presentations. However, those of interest to us are the computable presentations, which we define now.

Definition 3.24 Suppose $\mathcal{M}^\#$ is a presented metric structure. We say that $\mathcal{M}^\#$ is *computable* if it satisfies the following conditions:

1. There is an algorithm such that, given a code for a predicate symbol P, codes for generated points $p_1, \ldots, p_n$, and $k \in \mathbb{N}$, returns a rational number q such that $|P^{\mathcal{M}}(p_1, \ldots, p_n) - q| < 2^{-k}$.
2. The function $p \mapsto d_{\mathcal{M}}(p, M \setminus (S_n^{\mathcal{M}})^\circ)$ is lower semicomputable on the generated points of $\mathcal{M}^\#$ uniformly in n.[5]

An *index for* $\mathcal{M}^\#$ is then a natural number that codes these algorithms. We call $\mathcal{M}$ *computably presentable* if it admits a computable presentation.

Note, in particular, that since the metric symbol is always a distinguished predicate symbol, if $\mathcal{M}^\#$ is a computable presentation, then one can effectively approximate the distance on generated points of $\mathcal{M}^\#$ in each sort. As a result, *the metric is computable on the generated points of the presentation.*

The assumption (2) may seem unnecessary and unnatural, but we emphasize it is mainly for convenience. Like our earlier assumptions on sorts, the only proof in which it plays a significant role is that of Theorem 3.50. Note also that this assumption is satisfied by all of the examples considered in this and the ensuing chapters.

As examples of computable presentations, we note that the standard presentations of $\mathbb{F}^n$ and $C([0, 1]; \mathbb{F})$ are computable. If p is computable, then the standard presentations of ℓ^p, $L^p[0, 1]$, and ℓ_n^p are all computable. The standard presentation of ℓ_n^∞ is also computable.

As another example, suppose that $\mathcal{M}$ is a classical structure viewed as a metric structure equipped with the discrete metric. As mentioned above, there is essentially

[5] The term "lower semicomputable" is defined in Appendix B.

no difference between presentations of $\mathcal{M}$ as a classical structure and as a metric structure. It is a nice exercise for the reader to verify that such a presentation is computable in the classical sense if and only if it is computable in the continuous sense.

The following proposition shows that computable presentations are closed under products.

Proposition 3.25 *If $\mathcal{M}_j^\#$ is a computable presentation of $\mathcal{M}_j$ for each $j \in \{1, \ldots, n\}$, then $\mathcal{M}_1^\# \times \ldots \times \mathcal{M}_n^\#$ is a computable presentation of $\mathcal{M}_1 \times \ldots \times \mathcal{M}_n$. Moreover, an index for $\mathcal{M}_1^\# \times \ldots \times \mathcal{M}_n^\#$ can be computed from indices of the $\mathcal{M}_j^\#$'s.*

We usually cannot label all of the points of a metric structure with finite words formed from some fixed finite alphabet. However, if we allow for infinite words, then such a labeling becomes possible. These considerations motivate in part the concept of a name of a point, which we define as follows.

Definition 3.26 Suppose p is a point of a presented metric structure $\mathcal{M}^\#$. An $\mathcal{M}^\#$-*name* of p is a sequence $(p_n)_{n \in \mathbb{N}}$ of points of $\mathcal{M}^\#$ so that $d_{\mathcal{M}}(p, p_n) < 2^{-(n+1)}$ for all $n \in \mathbb{N}$.

Clearly, each point of a metric structure has a name, and no name names two points. Names have proven to be a fruitful device in computable analysis. In particular, as we shall see below, they provide a useful way to think about computability of functions on metric structures.

3.3 Computable Points and Sequences

With each presented metric structure (computable or not) there is associated a class of computable points.

Definition 3.27 Suppose $\mathcal{M}^\#$ is a presented metric structure and let p be a point of $\mathcal{M}$. We say that p is a *computable point of* $\mathcal{M}^\#$ if there is an algorithm that, given any $k \in \mathbb{N}$, computes a generated point ρ of $\mathcal{M}^\#$ so that $d(p, \rho) < 2^{-k}$.

It follows that every generated point of a presented metric structure $\mathcal{M}^\#$ is a computable point of $\mathcal{M}^\#$. If p is a computable point of $\mathcal{M}^\#$, then an $\mathcal{M}^\#$-*index* of p is an index of a Turing machine that computes p in the sense of Definition 3.27. We note also that if p is a point of a computably presented metric structure $\mathcal{M}^\#$, then p is a computable point of $\mathcal{M}^\#$ if and only if p has a computable $\mathcal{M}^\#$-name.

Example 3.28 Let $p \geq 1$ be a computable real number and let $(\ell^p)^\#$ be the presentation defined in Example 3.22. Then ζe_0 is a computable vector of $(\ell^p)^\#$. However, ζe_0 is not a computable vector of ℓ^p with respect to the standard presentation.

In the case of the standard presentation of $\mathbb{R}$, the computable points are simply the computable reals. There are two related but weaker concepts which we will use for forming certain counterexamples and stating some of our results. We define these now. Recall from Appendix B that $a \in \mathbb{R}$ is called *left c.e.* (respectively *right c.e.*) if $\mathbb{Q} \cap (-\infty, a)$ (respectively $\mathbb{Q} \cap (a, \infty)$) is c.e. and that a real number is computable if and only if it is both left and right c.e. The following example is due to Specker [12].

Example 3.29 Let $A \subseteq \mathbb{N}$ be c.e. and incomputable. Set $r = \sum_{n \in A} 2^{-n}$. It follows that r is left c.e. but not computable.

With each presented metric structure, there is associated a class of computable sequences.

Definition 3.30 Suppose $\mathcal{M}^\#$ is a presented metric structure and let $(p_n)_{n \in \mathbb{N}}$ be a sequence of points of $\mathcal{M}$. Then $(p_n)_{n \in \mathbb{N}}$ is a *computable sequence of* $\mathcal{M}^\#$ if there is an algorithm that, given $n, k \in \mathbb{N}$, computes a generated point ρ of $\mathcal{M}^\#$ so that $d_{\mathcal{M}}(p_n, \rho) < 2^{-k}$.

If $(p_n)_{n \in \mathbb{N}}$ is a computable sequence of $\mathcal{M}^\#$, then an $\mathcal{M}^\#$*-index* of $(p_n)_{n \in \mathbb{N}}$ is an index of a Turing machine that computes $(p_n)_{n \in \mathbb{N}}$ in the sense of Definition 3.30.

If $(p_n)_{n \in \mathbb{N}}$ is a computable sequence of $\mathcal{M}^\#$, then p_n is a computable point of $\mathcal{M}^\#$ for each $n \in \mathbb{N}$. However, if p_n is a computable point of $\mathcal{M}^\#$ for each $n \in \mathbb{N}$, it does *not* follow that $(p_n)_{n \in \mathbb{N}}$ is a computable sequence of $\mathcal{M}^\#$. For example, let A be an incomputable set of natural numbers, and for each $n \in \mathbb{N}$, let

$$p_n = \begin{cases} 1 & n \in A \\ 0 & n \notin A \end{cases}.$$

Then, since A is incomputable, it follows that $(p_n)_{n \in \mathbb{N}}$ is an incomputable sequence of $\mathbb{F}$. One can thus say that a computable sequence of $\mathcal{M}^\#$ is a sequence of "uniformly" computable points of $\mathcal{M}^\#$.

3.4 Topological Concepts

We now consider effective versions of three basic topological concepts: open set, closed set, and compact set. We only consider definitions and results directly related to developments in subsequent sections. For a more thorough exposition of computable topology, we refer the reader to [13]. Most of the results we state in this section are standard and have fairly straightforward proofs from the definitions. Hence, proofs will be omitted or sketched.

We begin by defining the rational balls associated with each presented metric structure.

Definition 3.31 Let $\mathcal{M}^\#$ be a presented metric structure. An open (respectively closed) ball B of $\mathcal{M}$ is a *rational open (respectively closed)* ball of $\mathcal{M}^\#$ if there exists a generated point ρ of $\mathcal{M}^\#$ and a positive rational number r so that $B = B(\rho; r)$ (respectively $B = \overline{B}(p; r)$).

We call each rational open ball of $\mathbb{R}$ a *rational open interval*. Each rational open ball of $\mathbb{R}^2$ or $\mathbb{C}$ is referred to as a *rational open disk*.

We can naturally extend the Gödel numbering of the generated points of $\mathcal{M}^\#$ to a Gödel numbering of the rational open (closed) balls of $\mathcal{M}$ that is effective in the sense that from a code of a rational open (closed) ball B it is possible to compute a code of a center for B as well as the radius of B.

In general, the inclusion relation for rational open balls is not necessarily decidable or even c.e. We are thus led to the following definition.

Definition 3.32 Suppose $B(\rho_0; r_0)$ and $B(\rho_1; r_1)$ are rational open balls of $\mathcal{M}_0^\#$ residing in the same sort. We say $B(\rho_0; r_0)$ is *formally included* in $B(\rho_1; r_1)$ if $d(\rho_0, \rho_1) + r_0 < r_1$.

If B and B' are rational open balls, we write $B \lhd B'$ if B is formally included in B'. For the sake of simplifying our discourse, we are obscuring the fact that formal inclusion is a relation on *codes* of rational open balls rather than on rational open balls. We note that the formal inclusion relation is not in general decidable, but at least it is c.e., which is often good enough for our purposes. We further note that if p is a point in an open rational ball B, then there is a rational open ball B' so that $p \in B'$ and $B' \lhd B$.

If $B(\rho; r)$ is a rational open ball of $\mathcal{M}^\#$, then we define the *formal diameter* of $B(\rho; r)$ to be $2r$. Again, for the sake of simplifying our exposition, we are ignoring the distinction between a rational ball and one of its codes. We denote the formal diameter of B by $\text{fdiam}(B)$.

With these definitions in hand, we can outline a proof of the following useful principle.

Proposition 3.33 *Suppose p is a point of a computable presentation $\mathcal{M}^\#$. Then p is a computable point of $\mathcal{M}^\#$ if and only if the set of rational open balls of $\mathcal{M}^\#$ that contain p is c.e.*

Proof To prove the proposition, let $\mathfrak{B}$ denote the set of all rational open balls of $\mathcal{M}^\#$ that contain p. First suppose that p is a computable point of $\mathcal{M}^\#$. For each $k \in \mathbb{N}$, compute a generated point ρ_k of $\mathcal{M}^\#$ so that $d_{\mathcal{M}}(\rho_k, p) < 2^{-k}$. Then for each rational ball B of $\mathcal{M}^\#$, we have $p \in B$ if and only if there exists $k \in \mathbb{N}$ so that $B(\rho_k; 2^{-k}) \lhd B$. Since the latter is a c.e. condition, we have that $\mathfrak{B}$ is c.e. Conversely, if $\mathfrak{B}$ is c.e., then given $k \in \mathbb{N}$, we search for $B \in \mathfrak{B}$ so that the formal diameter of B is less than 2^{-k}. The center of B is then a generated point within 2^{-k} of p. □

We now define the class of c.e. open sets associated with a presented metric structure by way of analogy with c.e. sets of natural numbers.

Definition 3.34 Suppose $\mathcal{M}^\#$ is a presented metric structure and suppose U is an open set of $\mathcal{M}$. We say that U is a *c.e. open set of* $\mathcal{M}^\#$ if it is the union of a c.e. set of rational open balls of $\mathcal{M}^\#$. In this case, by an $\mathcal{M}^\#$*-index* of U we mean an index of such a c.e. set of rational open balls.

Note that every rational open ball of a computable presentation $\mathcal{M}^\#$ is c.e. open. On the other hand, when a is a left-c.e. incomputable real, we have that (a, ∞) is not c.e. open.

The following analogs of the usual closure properties of open sets are direct consequences of the definitions.

Proposition 3.35 *Let $\mathcal{M}^\#$ be a computably presented structure.*

1. *If U, V are c.e. open sets of $\mathcal{M}^\#$, then $U \cap V$ is c.e. open.*
2. *If for each $n \in \mathbb{N}$, U_n is a c.e. open set of $\mathcal{M}^\#$ uniformly in n (that is, from n it is possible to compute an $\mathcal{M}^\#$-index of U_n), then $\bigcup_{n\in\mathbb{N}} U_n$ is c.e. open.*

We now discuss computability of closed sets. Specifically, with each computably presented metric structure there are associated classes of c.e. closed sets, co-c.e. closed sets, and computably closed sets.

Definition 3.36 Suppose $\mathcal{M}^\#$ is a presented metric structure and C is a closed set of $\mathcal{M}$.

1. C is a *c.e. closed set of* $\mathcal{M}^\#$ if the set of all rational open balls that intersect C is c.e. In this case, an $\mathcal{M}^\#$*-index of* C is an index of this set of rational open balls.
2. C is a a *co-c.e. closed set of* $\mathcal{M}^\#$ if its complement is a c.e. open set of $\mathcal{M}^\#$.
3. C is *computably closed* if C is c.e. closed and co-c.e. closed.
4. C is *strongly computably closed* if C is c.e. closed and if the set of all rational open balls included in $M \setminus C$ is c.e.

In order to develop some intuition for these concepts, it may be helpful to analogize a c.e. closed set to a c.e. set of natural numbers. Namely, when a set C is c.e.-closed, we can, by examining all rational open balls that contain a point of C, discern more and more information about which points lie in the set. Unlike a c.e. set of natural numbers, we can never say that a given point definitely lies in C, only that a nearby point lies in C without knowing which one.

The consideration of a few examples may also be helpful.

Example 3.37 By Proposition 3.33, $\{p\}$ is a c.e. closed set of $\mathcal{M}^\#$ if and only if p is a computable point of $\mathcal{M}^\#$.

Example 3.38 Let a be a left-c.e. incomputable real. Then $[a, \infty)$ is co-c.e. closed, but it is not c.e.-closed (and thus not computably closed). To see this, fix integers M, N so that $M < a < N$. For the sake of contradiction, suppose $[a, \infty)$ is c.e. closed. Fix $k \in \mathbb{N}$. Begin enumerating the rational intervals that intersect $[a, \infty)$. At the same time, enumerate the rational open intervals included in $(-\infty, a)$. Eventually, a finite set $\{I_1, \ldots, I_m, \ldots, I_{m+n}\}$ of rational intervals will be discovered that satisfies the following requirements:

- $[M, N] \subseteq \bigcup_j I_j$.
- The diameter of each I_j is smaller than 2^{-k}.
- Each I_j is included in $(-\infty, a)$ when $j \leq m$; otherwise $I_j \cap [a, \infty) \neq \emptyset$.

By inspecting this covering, we can discern j so that $a \in I_j$. It follows that a is computable, giving us a contradiction.

The following example is especially relevant to the discussion of word problems in Chapter 7.

Example 3.39 By means of a diagonalization argument, it is possible to construct a computably closed $C \subseteq \mathbb{R}$ that is not strongly computably closed (with respect to the standard presentation of $\mathbb{R}$). For each $e \in \mathbb{N}$, let $J_e = (2e, 2e + 1)$. At the beginning of our construction, we declare that every point in $\mathbb{R} \setminus \bigcup_e J_e$ belongs to C. Thus, we can list all rational open intervals that contain no point of $\bigcup_e J_e$ as containing a point of C.

At each stage s, we do the following for each $e \leq s$. If the eth Turing machine has not yet listed J_e as an interval that is included in $\mathbb{R} \setminus C$, then we declare that $(2e, 2e + 1 - 2^{-(s+1)})$ is included in the complement of C. Otherwise, declare $[2e + 1 - 2^{(-(s+1))}]$ to be included in C and begin listing all open rational intervals that intersect this closed interval as containing a point of C.

The following closure properties are straightforward consequences of the definitions.

Proposition 3.40 *Let $\mathcal{M}^\#$ be a computable presentation.*

1. *If $C_1, \ldots, C_n$ are c.e. closed sets of $\mathcal{M}^\#$, then $C_1 \cup \ldots \cup C_n$ is a c.e. closed set of $\mathcal{M}^\#$.*
2. *If $C_1, \ldots, C_n$ are co-c.e. closed sets of $\mathcal{M}^\#$, then $C_1 \cup \ldots \cup C_n$ is a co-c.e. closed set of $\mathcal{M}^\#$.*
3. *If for each $n \in \mathbb{N}$, C_n is a co-c.e. closed set of $\mathcal{M}^\#$ uniformly in n, then $\bigcap_{n \in \mathbb{N}} C_n$ is a co-c.e. closed set of $\mathcal{M}^\#$.*

Finally, we discuss the computability of compact sets. In fact, of the effective topological notions we discuss herein, that of computable compactness is often the most useful. Since our metric structures are by definition complete, their closed sets are compact if and only if they are totally bounded. Thus, we define computable compactness by effectivizing total boundedness.

Definition 3.41 Suppose $\mathcal{M}^\#$ is a presented metric structure and suppose C is a compact subset of $\mathcal{M}$. We say that C is a *computably compact set of* $\mathcal{M}^\#$ if there is an algorithm that, given any $k \in \mathbb{N}$, computes generated points $\rho_1, \ldots, \rho_n$ of $\mathcal{M}^\#$ so that $C \subseteq \bigcup_{j=1}^n B(\rho_j; 2^{-k})$ and so that $B(\rho_j; 2^{-k}) \cap C \neq \emptyset$ for each $j \in \{1, \ldots, n\}$. An index of such an algorithm is an $\mathcal{M}^\#$*-index of* C. We say that $\mathcal{M}^\#$ is *computably compact* if M is a computably compact set of $\mathcal{M}^\#$.

For example, every closed rational ball of $\mathbb{F}^n$ is a computably compact set of $\mathbb{F}^n$ with respect to its standard presentation. More generally, the following is proven in [14].

Theorem 3.42 *If $\mathcal{B}^\#$ is a computable presentation of a finite-dimensional Banach space, then the closed unit ball of $\mathcal{B}$ is a computably compact set of $\mathcal{B}^\#$.*

3.5 Computable Maps Between Presentations

With each pair $(\mathcal{M}_0^\#, \mathcal{M}_1^\#)$ of presented metric structures we associate a class of computable maps as follows.

Definition 3.43 Let $\mathcal{M}_0^\#$ and $\mathcal{M}_1^\#$ be presented metric structures, and consider a function $F : M_0 \to M_1$. We say F is a *computable map from $\mathcal{M}_0^\#$ to $\mathcal{M}_1^\#$* if there is an algorithm P that satisfies the following conditions:

1. Given a (code of an) open rational ball B_0 of $\mathcal{M}_0^\#$, P either does not halt or produces a (code of a) rational open ball B_1 of $\mathcal{M}_1^\#$ so that $F[B_0] \subseteq B_1$.
2. For every $p \in M_0$ and $k \in \mathbb{N}$, there is an open rational ball B of $\mathcal{M}_0^\#$ so that $p \in B$ and so that, when given B, P produces an open rational ball whose formal diameter is smaller than 2^{-k}.

We shall refer to 3.43.1 as the *correctness criterion*, and we shall refer to 3.43.2 as the *convergence criterion*.

Example 3.44 Addition and multiplication are computable maps from $\mathbb{F}^2$ to $\mathbb{F}$ (when both are given their standard presentations).

Example 3.45 If $r \in [0, 1]$ is computable, then $f \mapsto f(r)$ is a computable map from $C([0, 1]; \mathbb{F})$ to $\mathbb{F}$ (when both are given their standard presentations).

The first consequence of our definition is that the class of computable maps is closed under composition.

Proposition 3.46 *Suppose $\mathcal{M}_j^\#$ is a computable presentation of a metric structure for each $j \in \{0, 1, 2\}$. If S is a computable map from $\mathcal{M}_1^\#$ to $\mathcal{M}_2^\#$ and if T is a computable map from $\mathcal{M}_0^\#$ to $\mathcal{M}_1^\#$, then $S \circ T$ is a computable map from $\mathcal{M}_0^\#$ to $\mathcal{M}_2^\#$.*

Later, we will sketch a proof of Proposition 3.46. Because of this proposition, we may think of presented metric structures as forming a category in which the morphisms are the computable maps. In fact, the product of presentations defined in Sect. 3.2 is a product in this category.

It follows immediately from Definition 3.43 that every computable map from $\mathcal{M}_0^\#$ to $\mathcal{M}_1^\#$ is a continuous map from $\mathcal{M}_0$ to $\mathcal{M}_1$. Another fundamental consequence of the definition is the following.

Proposition 3.47 *Suppose $\mathcal{M}_0^\#$ and $\mathcal{M}_1^\#$ are computable presentations of metric structures and that F is a computable map from $\mathcal{M}_0^\#$ to $\mathcal{M}_1^\#$. Then F is computable on computable points, that is, $F(p)$ is a computable point of $\mathcal{M}_1^\#$ whenever p is a computable point of $\mathcal{M}_0^\#$.*

Proof To prove the proposition, fix an algorithm P for computing F in the sense of Definition 3.43. Suppose p is a computable point of $\mathcal{M}_0^\#$. By Proposition 3.33, there is an algorithm that enumerates all rational open balls of $\mathcal{M}_0^\#$ that contain p; begin running P on all of these balls in parallel. Whenever a rational open ball B of $\mathcal{M}_1^\#$ is output, begin enumerating all rational open balls B' of $\mathcal{M}_1^\#$ that formally include B. The result of this process is an enumeration of all rational open balls of $\mathcal{M}_1^\#$ that contain $F(p)$, whence $F(p)$ is a computable point of $\mathcal{M}_1^\#$ by Proposition 3.33 again. □

We note that the proof of Proposition 3.47 is uniform in that an $\mathcal{M}_1^\#$-index of $F(p)$ can be computed from an $\mathcal{M}_0$-index of p.

Example 3.48 Suppose $p \geq 1$ is a computable real. Then $f \mapsto f(0)$ is a computable map from ℓ^p to $\mathbb{F}$ when both are given their standard presentations. However, if $(\ell^p)^\#$ is the presentation in Example 3.22, then, in light of Proposition 3.47, $f \mapsto f(0)$ is *not* a computable map from $(\ell^p)^\#$ to $\mathbb{F}$.

The following principle is used extensively to demonstrate the computability of maps. It represents a partial converse to Proposition 3.47.

Lemma 3.49 *Suppose $\mathcal{M}_0^\#$ and $\mathcal{M}_1^\#$ are computably presented metric structures and $T : M_0 \to M_1$ is a function. Suppose the following:*

1. *T has a computable modulus of uniform continuity.*
2. *T is uniformly computable on the generated points of $\mathcal{M}_0^\#$, that is, there is an algorithm that, given $k \in \mathbb{N}$ and a (code of a) rational point ρ of $\mathcal{M}_0^\#$, produces a (code of a) generated point ρ' of $\mathcal{M}_1^\#$ so that $d_{\mathcal{M}_1}(T(\rho), \rho') < 2^{-k}$.*

Then T is computable.

Proof To prove the lemma, let $g : \mathbb{N} \to \mathbb{N}$ be a computable modulus of continuity for T. Without loss of generality, we may assume g is increasing. Let an open rational ball B of $\mathcal{M}_0^\#$ be given. We first compute the largest $k \in \mathbb{N}$ so that $\text{fdiam}(B) < 2^{-g(k)}$; if there is no such k then there is no output. Let ρ denote the center of B. Compute a generated point ρ' of $\mathcal{M}_1^\#$ so that $d_{\mathcal{M}_1}(\rho', T(\rho)) < 2^{-k}$ and output $B(\rho'; 2^{-k+1})$. This algorithm witnesses that T is computable. □

A consequence of (the proof of) Lemma 3.49 is the following.

Theorem 3.50 *Suppose $\mathcal{M}^\#$ is a computable presentation of $\mathcal{M}$. Then:*

1. *$d_\mathcal{M}$ is a computable map from $\mathcal{M}^\# \times \mathcal{M}^\#$ into $\mathbb{R}$.*
2. *Every n-ary functional $f : M^n \to \mathbb{F}$ of $\mathcal{M}$ is a computable map from $(\mathcal{M}^n)^\#$ to $\mathbb{F}$.*
3. *Every n-ary operation T of $\mathcal{M}$ is a computable map from $(\mathcal{M}^n)^\#$ to $\mathcal{M}$.*

Proof Suppose, for simplicity, that T is a unary operation of $\mathcal{M}$, that is, $T = F^\mathcal{M}$ for some unary function symbol F of L. Let a rational open ball $B = B(\rho; r)$ be given. We use condition 2 of Definition 3.24 to search for $n \in \mathbb{N}$ so that $d(\rho, M \setminus (S_n^\mathcal{M})^\circ) > r$. If no such n is found, then there is no output. However, since the sorts of $\mathcal{M}$ are strongly increasing (see remarks in Sect. 3.2), this search must terminate for sufficiently small r. We then proceed as in the proof of Lemma 3.49, but we now utilize the modulus of continuity for T on $S_n^\mathcal{M}$. Predicates are handled similarly. □

Theorem 3.51 *Suppose $\mathcal{M}_0^\#$ and $\mathcal{M}_1^\#$ are computable presentations of metric structures and $f : M_0 \to M_1$ is a function. Then the following are equivalent:*

1. *f is a computable map from $\mathcal{M}_0^\#$ to $\mathcal{M}_1^\#$.*
2. *There is an oracle Turing machine that, given an $\mathcal{M}_0^\#$-name of any point p of $\mathcal{M}_0$, computes a $\mathcal{M}_1^\#$-name of $f(p)$.*

We will give a reasonably detailed proof of Theorem 3.51. We first need a few preliminaries. As remarked above, even if $\mathcal{M}^\#$ is computable, the formal inclusion relation on the rational open balls of $\mathcal{M}^\#$, while c.e., is not in general decidable. However, if $d_\mathcal{M}(p, q) \notin \mathbb{Q}$ whenever p, q are generated points of $\mathcal{M}^\#$, then it *is* decidable. This observation leads us to consider the following lemma due to Moschovakis [15].

Lemma 3.52 *Suppose $(r_n)_{n\in\mathbb{N}}$ is a computable sequence of real numbers. Then for every open interval $I \subseteq \mathbb{R}$, there is a computable $\alpha \in I$ so that $\alpha r_n \notin \mathbb{Q}$ for all $n \in \mathbb{N}$.*

Our proof of Theorem 3.51 also requires the following technical concept related to the computation of maps.

Definition 3.53 Suppose P is an algorithm that maps (codes of) rational open balls of $\mathcal{M}_0^\#$ to (codes of) rational open balls of $\mathcal{M}_1^\#$. We say that P is *weakly monotonic* if, whenever P is defined on B_0 at stage s_0 and undefined on $B_1 \lhd B_0$ at s_0, then P is defined on B_1 and the formal diameter of $P(B_1)$ is no larger than the formal diameter of $P(B_0)$.

Our proof of Theorem 3.51 will use the following sufficient condition for a map to be computable:

Lemma 3.54 *Suppose $\mathcal{M}_0^\#$ and $\mathcal{M}_1^\#$ are computable and $\mathcal{M}_0^\#$ has a decidable formal inclusion relation. Let $f : M_0 \to M_1$. Suppose there is a c.e. set $\mathcal{S}$ with the following properties.*

1. *Each element of $\mathcal{S}$ is a pair (B, B') where B is a rational open ball of $\mathcal{M}_0^\#$ and B' is a rational open ball of $\mathcal{M}_1^\#$.*

2. *If* $(B, B') \in \mathcal{S}$, *then* $f[B] \subseteq B'$.
3. *For each* $k \in \mathbb{N}$ *and* $p \in M_0$, *there exists* $(B, B') \in \mathcal{S}$ *so that* $p \in B$ *and* $\mathrm{fdiam}(B') < 2^{-k}$.

Then f *is a computable map from* $\mathcal{M}_0^\#$ *to* $\mathcal{M}_1^\#$ *via a weakly monotonic algorithm.*

Proof We begin the proof of the lemma by noting that, without loss of generality, we may assume that if $(B, B') \in \mathcal{S}$ and $B_0 \lhd B$, then $(B_0, B') \in \mathcal{S}$. By (3), $\mathcal{S}$ is nonempty; since $\mathcal{S}$ is c.e., there is a computable $f : \mathbb{N} \to \mathbb{N}^2$ so that $\mathcal{S} = \mathrm{ran}(f)$. Set $\mathcal{S}_t = \{f(j) \ : \ j < t\}$.

We now define the desired algorithm P. Given a rational open ball B_0 of $\mathcal{M}_0^\#$, do the following:

Step 1. Wait for the least t so that $B_0 \in \mathrm{dom}(\mathcal{S}_t)$.
Step 2. Set $\mathcal{F} = \{B' \ : (B, B') \in \mathcal{S}_t$ for some B with $B_0 \lhd B\}$. Note that, since $\mathcal{M}_0^\#$ has a decidable formal inclusion relation, we can compute a canonical index of $\mathcal{F}$.
Step 3. Compute $B_0' \in \mathcal{F}$ so that the formal diameter of B_0' is minimal among the formal diameters of the balls in $\mathcal{F}$.
Step 4. Output B_0'.

We verify that P witnesses that f is computable. It follows from (2) that the correctness criterion is satisfied. We verify the convergence criterion as follows. Take $p \in M_0$ and $k \in \mathbb{N}$. By (3), there exists $(B_0, B_0') \in \mathcal{S}$ so that $p \in B_0$ and $\mathrm{fdiam}(B_0') < 2^{-k}$. Choose $s_0 \in \mathbb{N}$ so that $(B_0, B_0') \in \mathcal{S}_{s_0}$. Since $\mathcal{S}_{s_0}$ is finite, there is a rational open ball B_1 so that $p \in B_1 \lhd B_0$ and so that $B_1 \notin \mathrm{dom}(\mathcal{S}_{s_0})$. However, since $B_1 \lhd B_0$, $B_1 \in \mathrm{dom}(\mathcal{S})$; let s_1 be the least number so that $B_1 \in \mathrm{dom}(\mathcal{S}_{s_1})$. Then $s_1 > s_0$ and P is defined on B_1 at stage s_1. Let $B_1' = P(B_1)$. Since $B_1 \lhd B_0 \in \mathrm{dom}(\mathcal{S}_{s_0})$, it follows from 3 that $\mathrm{fdiam}(B_1') < 2^{-k}$.

Finally, we show that P is weakly monotonic. Suppose P is defined on B_0 at s_0 and undefined on $B_1 \lhd B_0$ at s_0. Let $B_0' = P(B_0)$. By the definition of P, $(B_0, B') \in \mathcal{S}$ for some B'. Hence, $(B_1, B') \in \mathcal{S}$. Let s_1 be the least number so that $B_1 \in \mathrm{dom}(\mathcal{S}_{s_1})$. It follows that $s_1 > s_0$ and P is defined on B_1 at s_1. It then also follows from the definition of P that the formal diameter of $P(B_1)$ is no larger than the formal diameter of $P(B_0)$. □

We are now ready to prove Theorem 3.51.

Proof (Theorem 3.51) We first show that (1) implies (2). Suppose the algorithm P computes f. Let $\Lambda = (p_j)_{j \in \mathbb{N}}$ be an $\mathcal{M}_0^\#$-name of p and take $k \in \mathbb{N}$. Run P on all B so that $B(p_n; 2^{-n}) \lhd B$ for some n. Wait for a B' whose formal diameter is at most 2^{-k} to be output. To see that such a ball exists, first note that, by the convergence criterion, P converges on a rational ball B_0 of $\mathcal{M}_0^\#$ so that $p \in B_0$ and $\mathrm{fdiam}(B) < 2^{-k}$. Set $B' = P(B_0)$. Since $p \in B(p_m; 2^{-m})$ for all $m \in \mathbb{N}$, there exists $n \in \mathbb{N}$ so that $B(p_n; 2^{-n}) \lhd B_0$ as desired. Define q_k to be the center of B'. If we set Λ' to be $(q_k)_{k \in \mathbb{N}}$, then Λ' is a $\mathcal{M}_1^\#$-name for $F(p)$.

Suppose now that (2) holds; we show that (1) also holds. By Lemma 3.52, there exists a computable $\alpha_0 \in (0, 1)$ so that $\alpha_0 d_{\mathcal{M}_0}(\rho, \rho') \notin \mathbb{Q}$ whenever ρ, ρ' are generated points of $\mathcal{M}_0^\#$. Let $\mathcal{N}_0$ denote the metric structure that is identical to $\mathcal{M}_0$ except that $d_{\mathcal{N}_0} = \alpha_0 d_{\mathcal{M}_0}$. Let $\mathcal{N}_0^\#$ be the presentation of $\mathcal{N}_0$ that has the same distinguished points as $\mathcal{M}_0^\#$. Since the metric of $\mathcal{N}_0$ is a computable rescaling of the metric of $\mathcal{M}_0$, it is possible to computably transform $\mathcal{N}_0^\#$-names into equivalent $\mathcal{M}_0^\#$-names. Thus, combined with assumption (2), we see that there is an oracle Turing machine **M** that, given any $\mathcal{N}_0^\#$-name of a $p \in M_0$, produces an $\mathcal{M}_1^\#$-name of $f(p)$.

We first show that f is a computable map from $\mathcal{N}_0^\#$ to $\mathcal{M}_1^\#$. To see this, for symmetry reasons, set $\mathcal{N}_1 = \mathcal{M}_1$. For $j = 0, 1$, we say that a sequence $(\rho_0, \ldots, \rho_k)$ of generated points of $\mathcal{N}_j^\#$ is a *partial* $\mathcal{N}_j^\#$*-name* if $d_{\mathcal{N}_j}(\rho_m, \rho_{m+1}) < 2^{-(m+1)}$ for all $m < k$. If $\hat{\Lambda} = (\rho_0, \ldots, \rho_k)$ is a partial $\mathcal{N}_0^\#$-name and $\hat{\Lambda}' = (\rho_0', \ldots, \rho_n')$ is a partial $\mathcal{N}_1^\#$-name, we write $\mathcal{T}_s(\hat{\Lambda}) = \hat{\Lambda}'$ if it is the case that when $\hat{\Lambda}$ is written on the oracle tape of **M**, $\hat{\Lambda}'$ is written on the output tape at stage s. If $\mathcal{T}_s(\hat{\Lambda}) = \hat{\Lambda}'$, we say that $(\hat{\Lambda}, \hat{\Lambda}')$ is a *witness*. If, in addition, B is a rational open ball of $\mathcal{N}_0^\#$ and B' is a rational open ball of $\mathcal{N}_1^\#$, we say that $(\hat{\Lambda}, \hat{\Lambda}')$ *witnesses* (B, B') if $B \lhd B_{\mathcal{N}_0}(\rho_k; 2^{-k})$ and $B_{\mathcal{N}_1}(\rho_n'; 2^{-n}) \lhd B'$. Let $\mathcal{S}$ denote the set of all (B, B') for which there is a witness. Since the set of partial $\mathcal{N}_j^\#$-names is c.e., it follows that $\mathcal{S}$ is c.e.

We now show that $\mathcal{S}$ satisfies the hypotheses of Lemma 3.54. To begin, suppose $(B, B') \in \mathcal{S}$ and let $p \in B$. Let $(\hat{\Lambda}, \hat{\Lambda}') = ((\rho_0, \ldots, \rho_k), (\rho_0', \ldots, \rho_n'))$ be a witness of (B, B'). Since $B \lhd B(\rho_k; 2^{-k})$, we have $d_{\mathcal{N}_0}(p, \rho_k) < 2^{-k}$. It then follows that there is an $\mathcal{N}_0^\#$-name Λ of p so that $\Lambda \supset \hat{\Lambda}$. Let $\Lambda' = M(\Lambda)$. Thus, Λ' is an $\mathcal{N}_1^\#$-name of $f(p)$. Then, since $\mathcal{T}_s(\hat{\Lambda}) = \hat{\Lambda}'$ for some $s \in \mathbb{N}$, $\Lambda' \supset \hat{\Lambda}'$. Hence, $f(p) \in B_{\mathcal{N}_1}(\rho_n'; 2^{-n}) \lhd B'$. It follows that $\mathcal{S}$ satisfies 3.54.(2).

Now we show that $\mathcal{S}$ satisfies 3.54.(3). Take $p \in M_0$ and $k \in \mathbb{N}$. Let $\Lambda = (\rho_j)_{j \in \mathbb{N}}$ be an $\mathcal{N}_0^\#$-name of p and let $\Lambda' = (\rho_j')_{j \in \mathbb{N}} = M(\Lambda)$. Since $M(\Lambda) = \Lambda'$, there exists $n_0 \in \mathbb{N}$ so that, for every $n \geq n_0$, there exists $s \in \mathbb{N}$ for which $\mathcal{T}_s(\Lambda \upharpoonright (n+1)) \supset \Lambda' \upharpoonright (k+3)$. Set:

$$\begin{aligned} B &= B(\rho_{n_0}; 2^{-n_0}) \\ B' &= B(\rho_{k+2}; 2^{-(k+2)}) \\ \hat{\Lambda} &= \Lambda \upharpoonright (n_0 + 1) \\ \hat{\Lambda}' &= \Lambda' \upharpoonright (k+3) \end{aligned}$$

Then $(\hat{\Lambda}, \hat{\Lambda}')$ is a witness for (B, B') whence $(B, B') \in \mathcal{S}$. At the same time, $p \in B$ since Λ is a name for p. By definition, $\mathrm{fdiam}(B') < 2^{-k}$. Thus, $\mathcal{S}$ satisfies 3.54.(3). Consequently, by Lemma 3.54, we have that f is a computable map from $\mathcal{N}_0^\#$ to $\mathcal{M}_1^\#$ via a weakly monotonic algorithm P.

It remains to show that f is a computable map from $\mathcal{M}_0^\#$ to $\mathcal{M}_1^\#$. Let Q be an algorithm so that $Q(B_{\mathcal{M}_0}(\rho; r)) = P(B_{\mathcal{N}_0}(\rho; r))$; we claim that Q witnesses that f is a computable map from $\mathcal{M}_0^\#$ to $\mathcal{M}_1^\#$. Since $B_{\mathcal{M}_0}(\rho; r) \subseteq B_{\mathcal{N}_0}(\rho; r)$, it follows that Q satisfies the correctness criterion. We now verify the convergence criterion. Take $k \in \mathbb{N}$. Then P is defined on some B_0 so that $p \in B_0$ and $\mathrm{fdiam}(P(B_0)) < 2^{-k}$. Let

$B_0' = P(B_0)$ and let s_0 denote the stage at which the computation of $P(B_0)$ halts. There exist a generated point ρ_1 of $\mathcal{N}_0^\#$ and a rational number r_1 so that

$$p \in B_{\mathcal{M}_0}(\rho_1; r_1) \subseteq B_{\mathcal{N}_0}(\rho_1; r_1) \lhd B_0$$

and so that $P(B_{\mathcal{N}_0}(\rho_1; r_1))$ is undefined at stage s_0. Let $B_1 = B_{\mathcal{N}_0}(\rho_1; r_1)$. Since P is weakly monotonic, $P(B_1)$ is defined; let $B_1' = P(B_1)$. Thus, $Q(B_{\mathcal{M}_0}(\rho_1; r_1))$ is defined and $Q(B_{\mathcal{M}_0}(\rho_1; r_1)) = B_1'$. Since P is weakly monotonic, $\mathrm{fdiam}(B_1') \leq \mathrm{fdiam}(B_0') < 2^{-k}$. Thus, Q satisfies the convergence criterion, finishing the proof of the theorem. □

Returning to an earlier point, we can now offer a proof of Proposition 3.46, which stated that computable maps are closed under composition.

Proof (Proposition 3.46) Suppose that S is a computable map from $\mathcal{M}_1^\#$ to $\mathcal{M}_2^\#$ via an algorithm P and T is a computable map from $\mathcal{M}_0^\#$ to $\mathcal{M}_1^\#$ via an algorithm Q. Note that the set $\mathcal{S}$ satisfies the hypotheses of Lemma 3.54, where $\mathcal{S}$ consists of all pairs (B, B'') that satisfy the following conditions:

- B is a rational open ball of $\mathcal{M}_0^\#$ and B'' is a rational open ball of $\mathcal{M}_2^\#$.
- P is defined on B.
- There is a rational open ball B' of $\mathcal{M}_1^\#$ so that $P(B) \lhd B'$ and so that $Q(B') = B''$.

We conclude that $S \circ T$ is a computable map from $\mathcal{M}_0^\#$ to $\mathcal{M}_2^\#$. □

We turn our attention now to the computable analogs of the usual topological principles concerning continuous functions.

Theorem 3.55 *Suppose $\mathcal{M}_0^\#$ and $\mathcal{M}_1^\#$ are computable presentations of metric structures and F is a computable map from $\mathcal{M}_0^\#$ into $\mathcal{M}_1^\#$.*

1. *If U is a c.e. open set of $\mathcal{M}_1^\#$, then $F^{-1}[U]$ is a c.e. open set of $\mathcal{M}_0^\#$.*
2. *If C is a co-c.e closed set of $\mathcal{M}_1^\#$, then $F^{-1}[C]$ is a co-c.e. closed set of $\mathcal{M}_0^\#$.*
3. *If C is a computably compact set of $\mathcal{M}_0^\#$, then $F[C]$ is a computably compact set of $\mathcal{M}_1^\#$.*

Proof To prove (1), suppose that U is a c.e. open set of $\mathcal{M}_1^\#$. Let $\mathfrak{B}'$ be a c.e. set of rational open balls of $\mathcal{M}_1^\#$ so that $U = \bigcup \mathfrak{B}'$. Let P be an algorithm that computes F. Define $\mathcal{B}$ to be the set of all rational open balls B of $\mathcal{M}_0^\#$ so that P maps B to a rational open ball that is formally included in an element of $\mathfrak{B}'$. Note that $\mathcal{B}$ is c.e. By definition, $\bigcup \mathcal{B} \subseteq F^{-1}[U]$. Let $p \in F^{-1}[U]$. Then $F(p) \in B'$ for some $B' \in \mathfrak{B}'$. There is a $\delta > 0$ so that $B_1' \lhd B'$ whenever B_1' is a rational open ball of $\mathcal{M}_1^\#$ so that $p \in B_1'$ and $\mathrm{fdiam}(B_1') < \delta$. It now follows from the convergence criterion that $\bigcup \mathfrak{B} = F^{-1}[U]$.

(2) follows immediately from (1). To prove (3), suppose that C is a computably compact set of $\mathcal{M}_0^\#$. Here is the algorithm for showing that $F[C]$ is a computably compact set of $\mathcal{M}_1^\#$. Upon input $k \in \mathbb{N}$, search for the following:

- An $n \in \mathbb{N}$ and generated points $\rho_0, \ldots, \rho_m$ of $\mathcal{M}_0^\#$ so that $C \subseteq \bigcup_j B(\rho_j; 2^{-n})$ and $B(\rho_j; 2^{-k}) \cap C \neq \emptyset$.
- Rational open balls $B_0, \ldots, B_m$ of $\mathcal{M}_0^\#$ so that $B(\rho_j; 2^{-n}) \lhd B_j$ for each $j \leq m$ and our algorithm for computing F halts on each B_j.
- Rational open balls $B_0', \ldots, B_m'$ of $\mathcal{M}_1^\#$ so that $F[B_j] \subseteq B_j'$ and $\mathrm{fdiam}(B_j') < 2^{-k}$ for each $j \leq m$.

The algorithm then outputs the centers of $B_1', \ldots, B_p'$. It is readily verified that this algorithm shows that $F[C]$ is a computably compact set of $\mathcal{M}_1^\#$. □

Corollary 3.56 *Suppose $\mathcal{M}^\#$ is a computable presentation of a metric structure and let F be a computable map from $\mathcal{M}^\#$ into $\mathbb{R}$. Then for every computably compact set C of $\mathcal{M}^\#$,* $\max F[C]$ *and* $\min F[C]$ *are computable reals.*

Proof It suffices to show that the maximum and minimum of a computably compact set $K \subseteq \mathbb{R}$ are computable. To see this, set $r = \max K$ and suppose it is required to compute a rational number q so that $|q - r| < 2^{-k}$. We can first compute rational numbers $q_1, \ldots, q_n$ so that $K \subseteq \bigcup_s (q_s - 2^{-k}, q_s + 2^{-k})$ *and* so that $(q_s - 2^{-k}, q_s + 2^{-k}) \cap K \neq \emptyset$ for each $s \in \{1, \ldots, n\}$. We can then set $q = \max\{q_s + 2^{-k} : s = 1, \ldots, n\}$. □

The following can be obtained by modifying the classical proof that a continuous function on a compact space is uniformly continuous.

Corollary 3.57 *Suppose $\mathcal{M}_0^\#$ and $\mathcal{M}_1^\#$ are computable presentations of metric structures and suppose F is a computable map from $\mathcal{M}_0^\#$ into $\mathcal{M}_1^\#$. Further suppose that $\mathcal{M}_0^\#$ is computably compact. Then there is a computable modulus of uniform continuity for F.*

3.6 Reconciling with Prior Approaches

We now take a moment to reconcile our notion of a computably presented metric structure with pre-existing notions in the literature. Let us consider, for example, the definition of a computable presentation of a (real) Banach space appearing in Melnikov's article [16]. First, one defines a presentation of a metric space (M, d) to be a dense sequence of points $(q_n)_{n \in \mathbb{N}}$; this agrees with our definition of a presentation of a metric space (viewed as a structure in the empty language). Such a presentation is said to be computable if one can compute the distances between the q_n's. Once again, this definition agrees with our definition. Melnikov then defines what it means for a function $f : M \to M$ to be computable with respect to the given presentation; his definition is essentially the second condition of Theorem 3.51.

All this being said, given a Banach space $\mathcal{B}$, Melnikov defines a computable presentation of $\mathcal{B}$ to be a computable presentation of $\mathcal{B}$ as a metric space such that addition is a computable map $\mathcal{B}^2 \to \mathcal{B}$ and scalar multiplication by each rational scalar is a computable map $\mathcal{B} \to \mathcal{B}$. It is clear from Theorem 3.50 that our presentations of Banach spaces satisfy this condition by taking the q_n's to be the generated points. Conversely, given a sequence $(q_n)_{n\in\mathbb{N}}$ that satisfies this condition, if we take the q_n's to be our distinguished points, it then follows that the distances from the generated points are uniformly computable, and so we have a computable presentation.

3.7 Adapting Classical Computable Structure Theory to Metric Structures

At this point, very little is required to extend classical computable structure theory to metric structures. As defined above, we define a metric structure to be *computably presentable* if it has a computable presentation. The property of computable presentability is preserved by isomorphism. Thus, it is an intrinsic property of the metric structures. The concepts of computable categoricity, degree of categoricity, computable dimension, and index set also naturally lift to the setting of metric structures without modification. As expected, computable dimension is preserved by isometric isomorphism.

The following observation, which is essentially due to Melnikov [16], is useful for transferring computable categoricity of a metric structure to an expansion.

Proposition 3.58 *Suppose $\mathcal{M}$ is a computably categorical metric structure and suppose that, for each $j \in \mathbb{N}$, T_j is an operation on $\mathcal{M}$ that is preserved by every isometric automorphism of $\mathcal{M}$. Then the expansion $(\mathcal{M}, (T_j)_{j\in\mathbb{N}})$ is computably categorical.*

Proof Set $\mathcal{N} = (\mathcal{M}, (T_j)_{j\in\mathbb{N}})$. Suppose $\mathcal{N}^\dagger$ and $\mathcal{N}^\#$ are computable presentations of $\mathcal{N}$. Let $\mathcal{M}^\dagger$ be the presentation of $\mathcal{M}$ whose distinguished points are precisely the generated points of $\mathcal{N}^\dagger$. Define $\mathcal{M}^\#$ similarly. Thus, every generated point of $\mathcal{M}^\dagger$ is a generated point of $\mathcal{N}^\dagger$ and similarly for $\mathcal{M}^\#$ and $\mathcal{N}^\#$. It immediately follows that $\mathcal{M}^\dagger$ and $\mathcal{M}^\#$ are computable, whence there is a computable isomorphism ϕ from $\mathcal{M}^\dagger$ onto $\mathcal{M}^\#$. By assumption, ϕ is also an isometric automorphism of $\mathcal{M}$. Finally, since every generated point of $\mathcal{M}^\dagger$ is a distinguished point of $\mathcal{N}^\dagger$ and every generated point of $\mathcal{M}^\#$ is a generated point of $\mathcal{N}^\#$, ϕ is also a computable map from $\mathcal{N}^\dagger$ onto $\mathcal{N}^\#$. □

3.8 Sorting It Out

In our discussion, we have assumed the sorts of our metric structures satisfy the assumptions introduced at the beginning of Sect. 3.2, which may seem stringent and confining even though they are satisfied by a wide array of examples. Here, we consider broadening this framework so as to remove these assumptions. In particular, we wish to assume only that our sorts are related by the inclusion maps and otherwise have no special relationship with each other. We also wish to dispose of condition 2 from Definition 3.24, which also seems ad hoc even though again it is satisfied by all the kinds of structures considered in subsequent chapters.

The most significant use of our assumptions on the sorts occurs in the proof sketch of Theorem 3.50. Once these are removed, Theorem 3.50 might then be rendered as follows.

Theorem 3.59 *Suppose that $\mathcal{M}^\#$ is a computable presentation. View each sort $S_i^{\mathcal{M}}$ as a computably presented metric space $S_i^{\mathcal{M}^\#}$ by taking as the set of distinguished points those generated points of $\mathcal{M}^\#$ that reside in S_i. Then:*

1. *If f is a function symbol with domain $S_{i_1} \times \cdots \times S_{i_n}$ and codomain S, then $f^{\mathcal{M}} : S_{i_1}^{\mathcal{M}^\#} \times \cdots \times S_{i_n}^{\mathcal{M}^\#} \to S^{\mathcal{M}^\#}$ is a computable function.*
2. *If P is a function symbol with domain $S_{i_1} \times \cdots \times S_{i_n}$ and codomain $[-R, R]$, then $P^{\mathcal{M}} : S_{i_1}^{\mathcal{M}^\#} \times \cdots \times S_{i_n}^{\mathcal{M}^\#} \to [-R, R]$ is a computable function.*

We note that the proof of this theorem now actually becomes even simpler, as the sort of an input rational open ball is part of the given data and does not need to be searched for. A similar amount of bookkeeping to keep track of the sorts would be applied to other definitions, theorems, and proofs. We spare the reader the details.

Alternatively, one may view each sort as a metric structure *in the signature of metric spaces*, that is, in a signature with no predicates or functions. We would then view a presentation of $\mathcal{M}$ as specifying a presentation for each sort. The computability of the presentation would then be determined by the computability of its functions as maps between the presentations of the sorts and the computability of the predicates as maps from the sorts to the reals.

References

1. Johanna N.Y. Franklin and Timothy H. McNicholl. Degrees of and lowness for isometric isomorphism. *J. Log. Anal.*, 12:Paper No. 6, 23, 2020.
2. Alec Fox. Computable presentations of C*-algebras. *J. Symb. Log.*, 89(3):1313–1338, 2024.
3. C. J. Ash and J. Knight. *Computable Structures and the Hyperarithmetical Hierarchy*, volume 144 of *Studies in Logic and the Foundations of Mathematics*. North-Holland Publishing Co., Amsterdam, 2000.
4. Antonio Montalbán. *Computable Structure Theory: Within the Arithmetic*. Perspectives in Logic. Cambridge University Press, Cambridge; Association for Symbolic Logic, Ithaca, NY, 2021.
5. A. Fröhlich and J. C. Shepherdson. Effective procedures in field theory. *Philos. Trans. Roy. Soc. London. Ser. A.*, 248:407–432, 1956.
6. Ekaterina B. Fokina, Iskander Kalimullin, and Russell Miller. Degrees of categoricity of computable structures. *Arch. Math. Logic*, 49(1):51–67, 2010.
7. Barbara F. Csima, Johanna N.Y. Franklin, and Richard A. Shore. Degrees of categoricity and the hyperarithmetic hierarchy. *Notre Dame J. Formal Logic*, 54(2):215–231, 2013.
8. Denis R. Hirschfeldt, Bakhadyr Khoussainov, Richard A. Shore, and Arkadii M. Slinko. Degree spectra and computable dimensions in algebraic structures. *Ann. Pure Appl. Logic*, 115(1–3):71–113, 2002.

9. Wesley Calvert. The isomorphism problem for classes of computable fields. *Arch. Math. Logic*, 43(3):327–336, 2004.
10. Wesley Calvert and Julia F. Knight. Classification from a computable viewpoint. *Bull. Symbolic Logic*, 12(2):191–218, 2006.
11. Martino Lupini. An invitation to model theory and C*-algebras. *Bull. Symb. Log.*, 25(1):34–100, 2019.
12. E. Specker. Nicht konstruktiv beweisbare Sätze der Analysis. *The Journal of Symbolic Logic*, 14:145–158, 1949.
13. Vasco Brattka and Peter Hertling, editors. *Handbook of Computability and Complexity in Analysis*. Theory and Applications of Computability. Springer, Cham, 2021.
14. Timothy H. McNicholl. Computing the exponent of a Lebesgue space. *J. Log. Anal.*, 12:Paper No. 7, 28, 2020.
15. Yiannis N. Moschovakis. *Descriptive Set Theory*, volume 155 of *Mathematical Surveys and Monographs*. American Mathematical Society, Providence, RI, second edition, 2009.
16. Alexander G. Melnikov. Computably isometric spaces. *J. Symbolic Logic*, 78(4):1055–1085, 2013.
17. Marian B. Pour-El and J. Ian Richards. *Computability in Analysis and Physics*. Perspectives in Mathematical Logic. Springer-Verlag, Berlin, 1989.

Chapter 4
Metric Spaces

We now shift our focus from metric structures in general to specific classes of these structures. We begin with the simplest of all metric structures, the metric spaces. In this case, the rational points and distinguished points of a presentation coincide. Since there are no operations, metric spaces are isomorphic if and only if they are isometric.

It is very easy to give examples of metric spaces that do not have a computable presentation: we simply take two points at an incomputable distance from each other. However, it is also easy to give examples of metric spaces that do have a computable presentation, and thus we begin our exploration with computable categoricity.

4.1 Computable Categoricity and Relative Categoricity

We first focus on one of the simplest kinds of metric spaces: spaces of the form (X, d) where $X \subseteq \mathbb{R}^n$ and d is the Euclidean metric. The completeness requirement for metric structures entails that X is closed. We then consider the Cantor space, $2^{\mathbb{N}}$, the Baire space, $\mathbb{N}^{\mathbb{N}}$, and the Urysohn space $\mathbb{U}$. All of these spaces are computably categorical. We then consider certain Banach spaces as metric spaces; some of these turn out *not* to be computably categorical. We close this section with some general results on computable categoricity and relative categoricity for metric spaces.

4.1.1 Subspaces of Euclidean Space

It is rather remarkable that there is an essentially geometric characterization of computable categoricity for these spaces. The following definition is due to Melnikov [1, Definition 8.1].

J. N. Y. Franklin et al., *Effective Metric Structure Theory*, Springer Monographs in Mathematics, https://doi.org/10.1007/978-3-032-13322-9_4

Definition 4.1 Let $M \subseteq \mathbb{R}^n$ be a closed subset whose affine span has dimension m. We say that $(x_0, \ldots, x_m)$ is an *intrinsically computable base of M* if it satisfies the following conditions:

1. The vectors $x_0 - x_i$, $1 \leq i \leq m$, are linearly independent in $\mathbb{R}^n$.
2. For every computable presentation $M^\#$ of M, there is a surjective self-isometry W of M such that each $W(x_i)$ is a computable point of $M^\#$.

In 2013, Melnikov proved the following [1, Theorem 8.2]:

Theorem 4.2 *Let $M \subseteq \mathbb{R}^n$ and suppose (M, d) is computably presentable. Then (M, d) is computably categorical if and only if M has an intrinsically computable base.*

Proof Let $\mathcal{M} = (M, d)$. We begin by assuming that $\mathcal{M}$ is computably categorical and fix a computable presentation $\mathcal{M}^\#$ of $\mathcal{M}$. It follows that there are distinguished points $x_0, \ldots, x_m$ of $\mathcal{M}^\#$ so that the vectors $x_0 - x_i$, $1 \leq i \leq m$ are linearly independent. That $(x_0, \ldots, x_m)$ is a computable base of $\mathcal{M}$ follows immediately from the computable categoricity of $\mathcal{M}$.

For the converse, we may suppose, without loss of generality, that the affine span of M is all of $\mathbb{R}^n$. (If it were not, we could isometrically map into a lower dimension and proceed.) Let $(x_0, \ldots, x_n)$ be an intrinsically computable base of M. Suppose that $\mathcal{M}^\#$ and $\mathcal{M}^\dagger$ are computable presentations of $\mathcal{M}$. Let $W^\#$ and $W^\dagger$ be surjective self-isometries of M such that, for each $i = 0, \ldots, n$, $W^\#(x_i)$ is $\mathcal{M}^\#$-computable and $W^\dagger(x_i)$ is $\mathcal{M}^\dagger$-computable. For each $i = 1, \ldots, n$, set $v_i = W^\#(x_i) - W^\#(x_i)$ and $w_i = W^\dagger(x_i) - W^\dagger(x_i)$. Let $(\mathbb{R}^n)^\#$ denote the presentation of $\mathbb{R}^n$ whose distinguished points consist of the linear span of $\{v_1, \ldots, v_n\}$ over $\mathbb{Q}$, and similarly let $(\mathbb{R}^n)^\dagger$ be the presentation of $\mathbb{R}^n$ whose distinguished points consist of the linear span of $\{w_1, \ldots, w_n\}$ over $\mathbb{Q}$. By Corollary 4.25 below (whose proof does not depend on the present result), $(\mathbb{R}^n)^\#$ and $(\mathbb{R}^n)^\dagger$ are computably isometric. Moreover, by the proof of Corollary 4.25, one may take the witnessing isomorphism Ψ to be the one that is given by $\Psi(v_i) = w_i$. Since Ψ takes M onto itself, the desired result follows. □

4.1.2 *Sequence Spaces: Cantor and Baire Space*

Another natural class of metric spaces to consider are the sequence spaces. Accordingly, we now turn our attention to Cantor space $2^{\mathbb{N}}$ and Baire space $\mathbb{N}^{\mathbb{N}}$. The metric for each of these spaces is defined by $d(\xi, \phi) = \max(\{2^{-n} \mid \xi(n) \neq \phi(n)\} \cup \{0\})$. A *cylinder* in one of these spaces is the set of all extensions of some finite sequence. We note that every sphere in Cantor space is a cylinder. We also note that every cylinder is open in Cantor space.

In Baire space, a sphere is not a cylinder, but it is a countable union of cylinders. Again, cylinders are open in Baire space, and so spheres are open in Baire space.

Each of these spaces has a standard presentation which is defined as follows. We first take an effective enumeration $(\sigma_j)_{j\in\mathbb{N}}$ of the corresponding finite sequence space (for example, in the case of Baire space, $\mathbb{N}^*$). We then define the jth distinguished point to be $\sigma_j{}^\frown 0^\omega$ (that is, σ_j followed by all zeros). It follows that these presentations are computable. In the case of Cantor space, it is useful to assume $(\sigma_j)_{j\in\mathbb{N}}$ enumerates 2^* in lexicographic order. We note that although this enumeration contains repetitions, it is decidable whether two distinguished points coincide.

Cantor space and Baire space were explored by Melnikov in [1] and Thewmorakot in [2]. Both of these spaces are computably categorical.

Theorem 4.3 *Cantor space* $2^{\mathbb{N}}$ *is computably categorical.*

Proof Set $X = 2^{\mathbb{N}}$. Let $X^\dagger$ be a computable presentation of X. We construct a computable map Φ from (the standard presentation of) X to $X^\dagger$ that is isometric and surjective. We first note that since X is compact, isometry implies surjectivity. Since Φ must be 1-Lipschitz, it suffices to define Φ on the distinguished points of X. That is, it suffices to define a computable map $\widehat{\Phi}$ on the distinguished points of X that is isometric, and then take Φ to be the continuous extension of $\widehat{\Phi}$ to X. In fact, we can do even better. Namely, we define $\widehat{\Phi}$ so that $\widehat{\Phi}(f)$ is a distinguished point of $X^\dagger$ whenever f is a distinguished point of X.

Let p_i denote the ith distinguished point of X, and let q_j denote the jth distinguished point of $X^\dagger$. The construction of $\widehat{\Phi}$ proceeds by stages. We let $\widehat{\Phi}_s$ denote the portion of $\widehat{\Phi}$ that has been defined by the end of stage s. At stage $s = 0$, we set $\widehat{\Phi}_0(p_0) = q_0$. Let $s \in \mathbb{N}$, and suppose $\widehat{\Phi}_s(p_k)$ is defined for each $k \leq s$. Let us say that $g \in X$ is *good for extension at* $s + 1$ if $\widehat{\Phi}_s \cup \{(p_{s+1}, g)\}$ is a partial isometry. We explain below why there is a distinguished point of $X^\dagger$ that is good for extension and why it can be found effectively. Taking all this for granted, we define $\widehat{\Phi}(p_{s+1})$ to be a distinguished point of $X^\dagger$ that is good for extension.

In order to finish the proof, we must demonstrate that there is a distinguished point of $X^\dagger$ that is good for extension. If $p_{s+1} = p_k$ for some $k \leq s$, then this is obvious, so we suppose $p_{s+1} \neq p_k$ for all $k \leq s$. We then seek q so that

$$q \in \bigcap_{k\leq s}\{g \in 2^{\mathbb{N}} \ : \ d(g, q_{f(k)}) = d(p_{s+1}, p_k)\}.$$

By the remarks above, each of the spheres in this intersection is a cylinder. Thus, this intersection is open. By an inductive argument, it is nonempty. (This is where the lexicographic ordering of the distinguished points is essential.) Since the distinguished points are dense, the existence of a good one follows. □

Now we turn our attention to Baire space.

Theorem 4.4 *Baire space is computably categorical.*

Proof sketch: Set $X = \mathbb{N}^{\mathbb{N}}$ and let $X^{\#}$ be a computable presentation of X. As in the proof of Theorem 4.3, we construct an isometry Φ by defining a computable and isometric map from the distinguished points of X to the distinguished points of $X^{\#}$. However, since Baire space is not compact, we must address the surjectivity requirement directly. We do so by ensuring that the range of $\widehat{\Phi}$ is dense.

Let p_i denote the ith distinguished point of X, and let q_j denote the jth distinguished point of $X^{\dagger}$.

Once again, our construction proceeds by stages. At stage $s = 0$, we define $\widehat{\Phi}_0(p_0) = q_0$.

Now let $s \in \mathbb{N}$. Assume we have defined $\widehat{\Phi}_s$ and that it is a partial isometry.

Suppose $s + 1 = 2n + 1$. We then ensure that $p_n \in \operatorname{dom}(\widehat{\Phi})$. If $p_n \in \operatorname{dom}(\widehat{\Phi}_s)$, we set $\widehat{\Phi}_{s+1} = \widehat{\Phi}_s$. Now we suppose $p_n \notin \operatorname{dom}(\widehat{\Phi}_s)$. We define ℓ to be the largest number such that $d(p, p_n) = 2^{-\ell}$ for some $p \in \widehat{\Phi}$. We then set $I = \{p \in \operatorname{dom}(\widehat{\Phi}_s) \mid d(p, p_n) = 2^{-\ell}\}$. In other words, I consists of the points in $\operatorname{dom}(\widehat{\Phi}_s)$ that are as close as possible to p_n. We then search for a distinguished point q' of $X^{\dagger}$ such that q' is not in the range of $\widehat{\Phi}_s$ and so that $d(\widehat{\Phi}_s(p), q') = 2^{-\ell}$ for all $p \in I$. We then set $\widehat{\Phi}_{s+1} = \widehat{\Phi}_s \cup \{(p_n, q')\}$. We will sketch the verification of this procedure below.

Now we suppose $s + 1 = 2n + 2$ and attend to showing that the range of $\widehat{\Phi}$ is dense. Let $B = B(q; r)$ denote the nth rational open ball of $X^{\dagger}$. We may assume $r = 2^{-k}$ for some k. Since distances between distinct points are powers of 2, we can then wait until it is witnessed that $\operatorname{ran}(\widehat{\Phi}_s) \cap B \neq \emptyset$ or it is witnessed that $\operatorname{ran}(\widehat{\Phi}_s) \cap B = \emptyset$. In the former case, we set $\widehat{\Phi}_{s+1} = \widehat{\Phi}_s$. In the latter case, we search for a distinguished point p' of X so that $d(p', p) = d(q, \widehat{\Phi}_s(p))$ for all $p \in \operatorname{dom}(\widehat{\Phi}_s)$. We then set $\widehat{\Phi}_{s+1} = \widehat{\Phi}_s \cup \{(p', q)\}$. The procedure for finding p' is similar to the procedure for finding q' when $s + 1$ is odd.

In order to verify the construction, we first consider the procedure at an odd stage $s + 1 = 2n + 1$. Since $\widehat{\Phi}_s$ is isometric, it follows from the definition of I that $\bigcap_{p \in I} B(\widehat{\Phi}_s(p); 2^{-\ell})$ is the union of countably many cylinders. Thus, this intersection is open, and we may again appeal to the density of the distinguished points for the existence of q'. The maximality of ℓ ensures that $\widehat{\Phi}_{s+1}$ is isometric. The verification of the procedure at an even stage is similar. □

4.1.3 The Urysohn Space

We now address the Urysohn metric space $\mathbb{U}$, which has been a source of great interest as of late, especially to descriptive set theorists [3, 4] and model theorists [5, 6]. The original construction can be found in [7, 8]; a comparison of three constructions can be found in [9]. It is characterized up to isometry by the following three properties:

- $\mathbb{U}$ is separable and complete.
- (Universality) $\mathbb{U}$ contains an isometric copy of every separable metric space.
- (Ultrahomogeneity) Every isometry between two finite subsets of $\mathbb{U}$ can be extended to an isometry of $\mathbb{U}$ onto itself.

It is worth noting that $\mathbb{U}$ is homeomorphic to a Hilbert space [4].

The Urysohn space has a computable presentation. To see this, one can introduce the *rational Urysohn metric space* $\mathbb{QU}$. As Melnikov points out in [1], $\mathbb{QU}$ is the Fraïssé limit of finite rational-valued metric spaces (in an appropriate classical first-order language). The completion of $\mathbb{QU}$ is isometric to $\mathbb{U}$. Since we can list all finite rational-valued metric spaces, the points in the rational Urysohn space form a natural computable presentation of the Urysohn space itself. We regard this as the standard presentation of $\mathbb{U}$. It is worth noting that the Urysohn space itself is the (continuous logic) Fraïsse limit of the class of finite metric spaces [10].

Interestingly, Urysohn's original construction of $\mathbb{U}$ is effective (see [9] for a treatment of this result). This allows one to prove the following [1, Theorem 7.3]:

Theorem 4.5 *The Urysohn space is computably categorical.*

In order to prove Theorem 4.5, we will need the following classical definition. For a metric space X, we say that a map $f : X \to \mathbb{R}^+$ is a *Katetov map* if, for any $y \notin X$, one can define a metric on $X \cup \{y\}$ extending the metric on X by defining $d(x, y) = f(x)$. We set $E(X)$ to be the the collection of all Katetov maps of X. We define a metric on $E(X)$ by setting $d(f, g) = \sup_{x \in X} |f(x) - g(x)|$.

We say that the metric space X satisfies the *approximate extension property* if, for every finite $A \subseteq X$, every $f \in E(A)$, and every $\epsilon > 0$, there is $x \in X$ so that $|d(x, a) - f(a)| \leq \epsilon$ for all $a \in A$. In other words, any abstract one-point extension of a finite subset of X can be approximately realized in X itself. It is a classical result that a Polish metric space has the approximate extension property if and only if it is isometric to the Urysohn space, which has the *exact extension property*. This is defined exactly as the approximate extension property but taking $\epsilon = 0$.

The proof of Theorem 4.5 is rather involved, so we will simply explain its components and refer the interested reader to [1] for all of the details.

Proof (Theorem 4.5) Suppose $\mathbb{U}^\#$ is a computable presentation of $\mathbb{U}$. Let p_j denote the jth distinguished point of $\mathbb{U}$, and let q_j denote the jth distinguished point of $\mathbb{U}^\#$. We construct two sequences $(\overline{p}_j)_{j\in\mathbb{N}}$ and $(\overline{q}_j)_{j\in\mathbb{N}}$ of points of $\mathbb{U}$ that satisfy the following conditions:

1. $(\overline{p}_j)_{j\in\mathbb{N}}$ is a computable sequence of $\mathbb{U}$ and $(\overline{q}_j)_{j\in\mathbb{N}}$ is a computable sequence of $\mathbb{U}^\#$.
2. $(\overline{p}_j)_{j\in\mathbb{N}}$ and $(\overline{q}_j)_{j\in\mathbb{N}}$ are dense in $\mathbb{U}$.
3. For all $j, k \in \mathbb{N}$, $d(\overline{p}_j, \overline{p}_k) = d(\overline{q}_j, \overline{q}_k)$.

It follows that there is a unique continuous extension of the map $\overline{p}_j \mapsto \overline{q}_j$. Furthermore, this extension is a computable map from $\mathbb{U}$ to $\mathbb{U}^\#$ that is isometric and surjective.

We define $\overline{p}_j$ by constructing a strongly Cauchy sequence $(\overline{p}_{j,s})_{s\in\mathbb{N}}$ of distinguished points of $\mathbb{U}$ and setting $\overline{p}_j = \lim_s \overline{p}_{j,s}$. We construct the $\overline{q}_j$'s similarly.

Suppose j is even. We define $\overline{p}_j$ so as to attend to the density requirement. We suppose $\overline{p}_i$ has been defined for each $i < j$. More precisely, we assume we have computed a $\mathbb{U}$-index of p_i for each $i < j$. We now search for $k, k' \in \mathbb{N}$ so that $k' > j$, $d(p_k, \overline{p}_i) > 2^{-k'}$ when $i < j$, and $\min\{d(p_m, \overline{p}_i) : i < j\} < 2^{-k'}$ when $m < k$. We define $\overline{p}_{j,s} = p_m$ for all $s \in \mathbb{N}$.

We now define $\overline{q}_j$. To begin, we search for a distinguished point q' of $\mathbb{U}^\#$ so that $|d(q', \overline{q}_i) - d(\overline{p}_j, \overline{p}_i)| < 1$ for all $i < j$. The existence of a point q' that satisfies this condition follows from the approximate extension property via a judicious choice of a Katetov map (see the proof of Theorem 3.4 of [3] for details). The existence of a distinguished point that satisfies this condition then follows from the density of the distinguished points. We then set $\overline{q}_{j,0} = q'$. Having defined $\overline{q}_{j,s}$, we define $\overline{q}_{j,s+1}$ as follows. We first search for a distinguished point q'' of $\mathbb{U}^\#$ so that $d(\overline{q}_{j,s}, q'') < 2^{2-s}$ and so that $|d(q'', \overline{q}_i) - d(\overline{p}_j, \overline{p}_i)| < 2^{-s}$ for all $i < j$. Again, the existence of such a point follows from the approximate extension property (and again we refer to the proof of Theorem 3.4 in [3] for details).

When j is odd, we define $\overline{p}_{j,s}$ and $\overline{q}_{j,s}$ as above, mutatis mutandis, so as to attend to the density of the $\overline{q}_j$'s. □

Given $r > 0$, one defines the *bounded Urysohn space of diameter* r to be the Fraïsse limit $\mathbb{U}_r$ of finite metric spaces of diameter bounded by r; it is characterized by the bounded analogs of the characteristic properties of $\mathbb{U}$. One may define $\mathbb{QU}_r$ analogously as the Fraïsse limit of the finite rational-valued metric spaces of diameter at most r. Thewmorakot [2] proved the following on the computable presentability and categoricity of these spaces.

Theorem 4.6 *Let r be a positive real number. Then $\mathbb{U}_r$ is computably presentable if and only if r is left-c.e. In this case, $\mathbb{U}_r$ is computably categorical.*

4.1.4 Banach Spaces as Metric Spaces

So far, all of the metric spaces considered have been computably categorical. We can obtain some more interesting examples by considering certain Banach spaces as metric spaces. We first enumerate some useful principles.

Some General Principles: Expansions of the Signature

We make use of the following classical result:

Theorem 4.7 (Mazur-Ulam Theorem) *Suppose $\mathcal{B}_0$ and $\mathcal{B}_1$ are* real *Banach spaces and let $\Phi : \mathcal{B}_0 \to \mathcal{B}_1$ be a surjective (not necessarily linear) isometry. Then there exists $v_0 \in \mathcal{B}_1$ so that $u \mapsto \Phi(u) - v_0$ is linear.*

The following result of Melnikov [1, Theorem 3.4] is a fairly straightforward application of the Mazur-Ulam Theorem.

Proposition 4.8 *If a real Banach space is computably categorical as a metric space, then it is computably categorical.*

Once we move to complex Banach spaces, the Mazur-Ulam Theorem no longer applies. We are thus led to the following question.

Question 4.9 If a *complex* Banach space is computably categorical as a metric space, does it follow that it is computably categorical?

The following fact, pointed out in [1, Fact 3.6], also follows quickly from the Mazur-Ulam Theorem. It concerns the possibility of expanding a computable presentation of the underlying metric space of a Banach space $\mathcal{B}$ to a computable presentation of $\mathcal{B}$.

Proposition 4.10 *Let $\mathcal{B}$ be a real Banach space. Suppose $\mathcal{B}^{\#}$ is a computable presentation of $\mathcal{B}$, and suppose $\mathcal{B}^{\dagger}$ is a computable presentation of $\mathcal{B}$* as a metric space. *Further suppose that* $\mathbf{0}$ *is a computable point of $\mathcal{B}^{\dagger}$ and that $\mathcal{B}^{\#}$ and $\mathcal{B}^{\dagger}$ are computably isometric (as metric spaces). Then addition and scalar multiplication are uniformly computable with respect to $\mathcal{B}^{\dagger}$, and so $\mathcal{B}^{\dagger}$ can also be viewed as a computable presentation of $\mathcal{B}$ as a Banach space.*

The following observation is due to Diamondstone, but was first pointed out in [11, Facts 2.10 and 2.11].

Proposition 4.11 *Suppose that $\mathcal{B}$ is a real Banach space and $\mathcal{B}^{\#}$ is a computable presentation of $\mathcal{B}$ as a metric space for which addition or subtraction is a computable operation. Then for each rational number q, multiplication by q is a computable operation of $\mathcal{B}^{\#}$ uniformly in q. Thus, $\mathcal{B}^{\#}$ can be regarded as a computable presentation of $\mathcal{B}$ as a Banach space.*

Melnikov and Ng added the following observation specifically about $C([0, 1]; \mathbb{R})$ as a Banach algebra [11, Fact 2.15].

Proposition 4.12 *Let $\mathcal{B} = C([0, 1]; \mathbb{R})$, and suppose $\mathcal{B}^{\#}$ is a computable presentation of $\mathcal{B}$ as a metric space for which addition and multiplication are computable. Then for each rational number q, multiplication by q is a computable operation of $\mathcal{B}^{\#}$ uniformly in q. In addition, the constant function* 1 *is a computable point of $\mathcal{B}^{\#}$. Thus, $\mathcal{B}^{\#}$ may be regarded as a computable presentation of the Banach algebra $C([0, 1]; \mathbb{R})$ for which the unit is a computable point.*

$C([0, 1]; \mathbb{R})$ as a Metric Space

We now consider the Banach space $C([0, 1]; \mathbb{R})$ as a metric space. This is our first example of a metric space that is not computably categorical. We also demonstrate the existence of presentations of $C([0,1];\mathbb{R})$ in which various operations external to the signature of metric spaces fail to be computable. Finally, we demonstrate that as a metric space, $C([0,1];\mathbb{R})$ has computable dimension $\aleph_0$.

In this section, we abbreviate $C([0,1];\mathbb{R})$ by $C[0,1]$. We define the standard presentation of $C[0,1]$ (as a metric space) by defining the distinguished points to be the rational piecewise linear continuous functions. We begin with the following theorem of Melnikov [1, Theorem 5.2].

Theorem 4.13 *There is a computable presentation of* $C[0,1]$ *as a metric space in which* $\mathbf{0}$ *is a computable point but the operation of multiplying by* $\frac{1}{2}$ *is not computable.*

Proof This is proved by a diagonalization argument. Let ℓ_j denote the jth distinguished point of $C[0,1]$. We will produce a presentation $C[0,1]^{\#} = (C[0,1], (f_i)_{i\in\mathbb{N}})$ such that $f_i = \lim_s f_{i,s}$ where $(f_{i,s})_{i,s\in\mathbb{N}}$ is a computable double sequence of ℓ_j's. Thus, $(f_i)_{i\in\mathbb{N}}$ is a Δ^0_2 sequence of $C[0,1]$. Rather than give the entire proof, we will explain each of the three types of requirements and how they fit together. Let us suppose that at stage s, we have a finite set of functions $f_{0,s}, \ldots, f_{n(s),s}$ as described above for some function $n(s)$ that is nondecreasing in s. Set $\lambda(x) = 2^{-1}x$ for all $x \in [0,1]$.

To ensure that the zero vector $\mathbf{0}$ is a computable point, we guarantee in our construction that $\lim_s f_{0,s} = \mathbf{0}$.

At each stage $s+1$, we ensure that $d(f_{i,s}, f_{j,s}) = d(f_{i,s+1}, f_{j,s+1})$ for every i and j. Thus, $(d(f_i, f_j))_{i,j\in\mathbb{N}}$ is computable.

In order to ensure that $\lim_s f_{j,s}$ exists, we guarantee that for every n, there is a stage s such that for all $t, z > s$ and all i, we have $d(f_{i,t}, f_{i,z}) < 2^{-n}$. These requirements will not conflict with any others.

To ensure the density of $(f_i)_{i\in\mathbb{N}}$, for each j we ensure that ℓ_j is in the closure of $(f_i)_{i\in\mathbb{N}}$. To handle this requirement, we check at stages of the form $s = 2\langle k, j\rangle$ to see whether ℓ_j is among $f_{0,s}, \ldots, f_{n(s),s}$. If this is not so, we set $f_{n(s)+1,t} = \ell_j$ for every $t \le s+1$. Since we preserve the distance between our f_is as the construction progresses, this is enough.

Finally, for each $e \in \mathbb{N}$, we have the requirement that e is not a $C[0,1]^{\#}$-index of λ.

The strategy for this requirement will work inside its own interval $I_e = (a_e, b_e)$. On this interval, we define a continuous function w_e by

$$w_e(x) = \begin{cases} 0 & x \notin (a_e, b_e) \\ 2^{-e} & x \in [a_e + \frac{1}{4}(b_e - a_e), a_e + \frac{3}{4}(b_e - a_e)] \\ \text{linear} & \text{otherwise.} \end{cases}$$

We note that our w_e is constantly 2^{-e} in the "middle half" of the interval I_e and that this line segment is connected linearly to the line segment(s) that is (are) 0 outside of I_e. In short, we can use them to create a small bump in our function inside of I_e.

At stage $t = 2e+1$, we will set $f_{n(t)+1,r} = w_e$ for every $r \le t$ if w_e is not already one of $f_{0,t}, \ldots, f_{n(t),t}$. At $s > t$, if we have $p \le n(t)+1$ such that $f_{p,t} = w_e$, we look for a computation by the eth Turing machine that tells us that $d(f_h, \frac{1}{2}f_p) < 2^{-k}$ for some sufficiently large k. There are now three cases.

- If $f_{h,s}$ is not yet defined, we set $f_{h,v}$ for every $v \leq s$ to be a new rational piecewise linear function g so that g is very far away from $\frac{1}{2}f_{p,s}$ somewhere on I_e.
- If $f_{h,s}$ is already defined and $f_{h,s}$ is sufficiently far from $\frac{1}{2}f_{p,s}$ somewhere on I_e, we do nothing.
- If $f_{h,s}$ is already defined and sufficiently close to $\frac{1}{2}f_{p,s}$ on I_e, we define $f_{h,s+1}$ to be a function that is sufficiently close to $f_{h,s}$ but that "goes wrong" within some subinterval of I_e.

We can then verify that all the requirements are met and check that all the stages are effective. □

Since multiplication by $\frac{1}{2}$ is a computable operation of the standard presentation of the metric space $C[0, 1]$, we now have the following.

Corollary 4.14 $C[0, 1]$ *is not computably categorical as a metric space.*

Melnikov and Ng [11, Theorem 3.12] significantly strengthened this corollary.

Theorem 4.15 *The metric space* $C[0, 1]$ *has computable dimension* $\aleph_0$.

To arrive at this result, we need to consider limit equivalent presentations.

Definition 4.16 Suppose that $\mathcal{M}^{\#} = (M, (p_i))$ and $\mathcal{M}^{\dagger} = (M, (q_i))$ are two computable presentation of a metric space $\mathcal{M}$. Then we say that $\mathcal{M}^{\#}$ and $\mathcal{M}^{\dagger}$ are *limit equivalent* if there is a Δ_2^0-computable function $f : \mathbb{N} \to \mathbb{N}$ so that $p_j \mapsto q_{f(j)}$ is an isometric bijection.

Furthermore, we say that a computable presentation $\mathcal{M}^{\#}$ of a metric space $\mathcal{M}$ is *rational valued* if the distance between any two of its distinguished points is rational.

Melnikov and Ng adapted the proof of Theorem 4.13 to show the following:

Theorem 4.17 *There is a rational-valued computable presentation* $C[0, 1]^{\#}$ *of the metric space* $C[0, 1]$ *that is limit equivalent to the standard presentation such that the constant function* 0 *is a computable point of* $C[0, 1]^{\#}$ *but scalar multiplication by* $\frac{1}{2}$ *is not a computable operation of* $C[0, 1]^{\#}$.

Proof The proof of this theorem is very much like that of Theorem 4.13, but we add a requirement to make $C[0, 1]^{\#}$ limit equivalent to the standard presentation. The $(f_i)_{i\in\mathbb{N}}$'s from the previous proof may not be limit equivalent to the $(\ell_i)_{i\in\mathbb{N}}$'s because our interpretation of each $f_{i,s}$ may be changed infinitely often, though to a smaller and smaller extent each time. In order to change each of our elements only finitely many times, we must establish a witness piecewise linear function that has a constant value on some unique interval. If the functional in question converges with high accuracy, we then adjust our witnessing function by lowering its value at some point within this interval. While this does mean we have to adjust our other functions to maintain distance preservation, we will be able to do this while guaranteeing that functions with a low enough norm will not be adjusted after some stage and thus that the new presentation will be limit equivalent to the standard presentation. □

The following theorem of Melnikov and Ng [11], combined with Theorem 4.17, implies Theorem 4.15.

Theorem 4.18 *If a metric space $\mathcal{M}$ has two computable, rational-valued limit equivalent but not computably isometric presentations, then $\mathcal{M}$ has computable dimension $\aleph_0$.*

Finally, we immediately obtain the following by combining Proposition 4.11 and Theorem 4.13.

Corollary 4.19 *[11] There is a computable presentation of $C[0, 1]$ as a metric space for which addition is not computable.*

$C(2^{\mathbb{N}}; \mathbb{R})$ as a Metric Space

In this section, we study the metric space $C(2^{\mathbb{N}}; \mathbb{R})$, which consists of all continuous functions from $2^{\mathbb{N}}$ into $\mathbb{R}$. *Again, in this section, we abbreviate this space by $C(2^{\mathbb{N}})$.* The metric is the supremum metric. In his Ph.D. thesis [2], Thewmorakot proved the equivalent of Theorem 4.17 for this space.

Theorem 4.20 *There is a rational-valued presentation $C(2^{\mathbb{N}})^{\#}$ of the metric space $C(2^{\mathbb{N}})$ that satisfies the following:*

1. *$C(2^{\mathbb{N}})^{\#}$ is limit equivalent to the standard presentation of $C(2^{\mathbb{N}})$.*
2. *The constant function 0 is a computable point of $C(2^{\mathbb{N}})^{\#}$.*
3. *Scalar multiplication by $\frac{1}{2}$ is not a computable operation of $C(2^{\mathbb{N}})^{\#}$.*

The proof is very similar to the proof of Theorem 4.17. In this case, the counterparts of Melnikov and Ng's intervals are basic open sets.

We have the following corollaries.

Corollary 4.21 *$C(2^{\mathbb{N}})$ has computable dimension $\aleph_0$.*

Corollary 4.22 *There is a computable presentation of $C(2^{\mathbb{N}})$ as a metric space for which addition is not computable.*

Hilbert Spaces as Metric Spaces

Melnikov [1, Theorem 4.7] proved that real Hilbert spaces are computably categorical as metric spaces.

Theorem 4.23 *Every real Hilbert space is computably categorical as a metric space.*

To prove Theorem 4.23, we will need the following lemma of Melnikov [1, Lemma 4.1].

Lemma 4.24 *Suppose that $\mathcal{H}$ is a real Hilbert space and $\mathcal{H}^\#$ is a computable presentation of $\mathcal{H}$ as a metric space for which the vector $\mathbf{0}$ is a computable point. Then the addition and scalar multiplication operations are computable operations of $\mathcal{H}^\#$.*

Proof To prove the lemma, one first uses the parallelogram law and polarization identity to show that addition is computable with respect to $\mathcal{H}^\#$. It is then straightforward to show that multiplication by (-1) is a computable map, and from there the computability of multiplication by all rational scalars follows. □

Proof (Theorem 4.23) Suppose that $\mathcal{H}^\#$ is a computable presentation of a Hilbert space $\mathcal{H}$ as a metric space. While $\mathbf{0}$ may not be a computable point of $H^\#$, we can "reset" the value of $\mathbf{0}$ and adjust the operations accordingly. We declare this new zero vector to be p for some fixed distinguished point p of $\mathcal{H}^\#$. We then define the new vector space operations $+_1$ and $(r\cdot_1)$ that "recenter" our space around p:

$$x +_1 y = x + y - p$$
$$r \cdot_1 x = r \cdot (x - p) + p$$

Now, if we set $||x||_1 = d(p, x)$, we have an operation that satisfies the norm axioms with respect to these new operations as well as the parallelogram law. Hence, this space also has an inner product. By Lemma 4.24, all of these operations are computable operations of $\mathcal{H}^\#$. We can then produce a computable and linearly independent sequence of $H^\#$ whose linear span is dense. From this, we can produce an orthonormal sequence via Gram-Schmidt orthonormalization. It is then easy to define a computable linear map from $\mathcal{H}^\#$ to $\mathcal{H}$ that is isometric and surjective. □

Corollary 4.25 $\mathbb{R}^n$ *is computably categorical as a metric space.*

4.1.5 A Sufficient Condition

We take a moment to discuss a sufficient condition for the computable categoricity of a Polish space. Classically, a collection of points of a metric space is an *automorphism base* if the global behavior of a nontrivial surjective self-isometry of this space is completely determined by its behavior on the base. Melnikov effectivized this notion as follows [1]:

Definition 4.26 A finite automorphism base $B = \{b_1, \ldots, b_n\}$ of a Polish space $\mathcal{M} = (M, d)$ is an *effective automorphism base of* $\mathcal{M}$ if it satisfies the following conditions.

1. For every computable presentation $\mathcal{M}^{\#}$ of $\mathcal{M}$, there is a surjective self-isometry U of $\mathcal{M}$ such that each $U(b_i)$ is $\mathcal{M}^{\#}$-computable.
2. For every rational $\epsilon > 0$, we can compute $\delta \in (0, 1)$ such that, for every $x, y \in M$, we have that $d(x, y) < \epsilon$ whenever $\max_{i=1,\ldots,n} |d(b_i, x) - d(b_i, y)| < \frac{\delta}{C_{x,y}}$ where $C_{x,y}$ is defined by

$$C_{x,y} = 1 + \max_{i=1,\ldots,n} d(b_i, x) + \max_{i=1,\ldots,n} d(b_i, y).$$

In short, if B is an effective automorphism base of M, the elements of M are effectively determined by their distances to the points in B. In [1, Theorem 9.2], Melnikov proved that the existence of an effective automorphism base suffices to prove that a computably presentable structure is computably categorical:

Theorem 4.27 *Suppose a Polish space $\mathcal{M}$ has a computable structure. If $\mathcal{M}$ has an effective automorphism base, then $\mathcal{M}$ is computably categorical.*

Proof To prove the theorem, take an effective automorphism base $\{b_1, \ldots, b_n\}$ for $\mathcal{M}$ and two computable structures $\mathcal{M}^{\#} = (\mathcal{M}, (p_i)_{i\in\mathbb{N}})$ and $\mathcal{M}^{\dagger} = (\mathcal{M}, (q_i)_{i\in\mathbb{N}})$ on $\mathcal{M}$. We can then take two surjective self-isometries W_1 and W_2 such that $W_1(b_1), \ldots, W_1(b_n)$ and $W_2(b_1), \ldots, W_2(b_n)$ are computable with respect to $\mathcal{M}^{\#}$ and $\mathcal{M}^{\dagger}$, respectively. We define our desired surjective self-isometry U to be $W_2 \circ W_1^{-1}$; this will be a computable map from $\mathcal{M}^{\#}$ to $\mathcal{M}^{\dagger}$ by the defining properties of an effective automorphism base. □

As an example of this method, one can show that the unit circle $\mathbb{S}^1$ (where the distance between two points is defined to be the length of the shortest arc between them) is computably categorical: take a computable structure $(\mathbb{S}^1)^{\#} = (\mathbb{S}^1, (p_i)_{i\in\mathbb{N}})$ on $\mathbb{S}^1$ and observe that there must be p_j and p_k such that the distance between them is less than $\frac{1}{2}$. These points form an effective automorphism base of $\mathbb{S}^1$ [1].

Finally, Melnikov showed that, for closed subsets of Euclidean space, the existence of an automorphism base is a necessary and sufficient condition [1, Theorem 9.5].

Theorem 4.28 *If $M \subseteq \mathbb{R}^n$ is computably presentable, then the following are equivalent.*

1. *M is computably categorical.*
2. *M has an intrinsically computable base.*
3. *M has an effective automorphism base.*

4.1.6 Relative Categoricity of Compact Spaces

We mention in passing that Melnikov and Nies proved a general categoricity result for arbitrary compact metric spaces [12].

Theorem 4.29 *Every computably presentable compact metric space is Δ^0_3-categorical.*

Melnikov and Nies also showed that this bound cannot be improved to be Δ^0_2-categorical (see [13]).

4.2 Index Sets

In this section, we present a number of results regarding the complexity of index sets of computably presentable metric spaces. Except for the results at the end of the section, the results presented here are from Thewmorakot's thesis [2]. We begin with the complexity of all metric space presentations.

Theorem 4.30 *The set of all indices of metric space presentations is Π^0_2-complete.*

Discerning copies of individual metric spaces turns out to have fairly low complexity, as do some related problems.

Theorem 4.31 *If X is a computably presentable Polish metric space, then the following index sets are Π^0_2.*

1. *The set of all metric spaces that are isometric to X.*
2. *The set of all metric spaces that isometrically embed in X.*
3. *The set of all metric spaces into which X isometrically embeds.*

Theorem 4.31 follows quickly from the following more general result.

Theorem 4.32 *If $\mathcal{K}$ is a nonempty class of computably presentable Polish metric spaces, then the index set of $\mathcal{K}$ is Π^0_2.*

We now turn to the isometry problem.

Theorem 4.33 *The isometry problem for metric spaces is Σ^1_1-complete.*

Proof Let I be the set of all pairs (e, e') so that e and e' index presentations of isometric metric spaces. It is fairly obvious that I is Σ^1_1, since this allows one to search Baire space for a representation of an isometry. We now turn to the completeness. We begin by recalling the fact that there is a computable sequence $(H_i)_{i \in \mathbb{N}}$ of directed graphs such that the isomorphism problem for this sequence is Σ^1_1-complete [14]. We then use the coding technique in [15] to obtain a computable sequence $(G_i)_{i \in \mathbb{N}}$ of connected undirected graphs such that the set $J = \{(i, j) \in \mathbb{N}^2 \mid G_i \cong G_j\}$ is Σ^1_1-complete. We conclude by showing that J is many-one reducible to I. To see this, we regard any countable graph as a Polish metric space with the shortest path metric. It is straightforward to argue that G_i and G_j are isomorphic as graphs if and only if they are isometric as metric spaces. □

The embedding problem is similarly hard.

Theorem 4.34 *The embedding problem for metric spaces is Σ^1_1-complete.*

Proof Let I denote the set of all pairs (e_0, e_1) so that e_0 and e_1 index presentations of metric spaces M_0 and M_1 respectively and so that M_0 isometrically embeds in M_1. Again, the upper bound is fairly straightforward. Let $d_{\mathbb{N}}$ denote the usual metric on $\mathbb{N}$, and let J denote the set of all primitive recursive indices of subtrees of $\mathbb{N}^*$ with an infinite branch. It is well known that J is Σ^1_1-complete (see, for example, [16]). For each $e \in \mathbb{N}$, we compute a pseudometric d_e on $\mathbb{N}$ so that $e \in J$ if and only if $(\mathbb{N}, d_{\mathbb{N}})$ isometrically embeds into the completion of the quotient of $(\mathbb{N}, d_e)$. Intuitively, an infinite branch of a tree indexed by e represents such an isometric embedding. □

We also briefly mention Thewmorakot's work on finite and infinite metric spaces when considered separately. It turns out that frequently we have completeness results in these contexts and that often these problems are as difficult as they could reasonably be. Throughout the rest of this section, the variable M ranges over metric spaces only.

Theorem 4.35 *Suppose X is a computably presentable finite metric space.*

1. *If $|X| = 1$, then the index set of $\{X\}$ is Π^0_2-complete.*
2. *If $|X| > 1$, the index set of $\{M : X \hookrightarrow M\}$ is Σ^1_1-complete.*

Theorem 4.36 *Suppose X is an unbounded subset of $\mathbb{R}$. If (X, d) is computably presentable, then the index set of $\{M : X \hookrightarrow M\}$ is Σ^1_1-complete.*

Thewmorakot also proves that the following sets are Σ^1_1-complete:

- The index set of the set of all metric spaces into which $2^{\mathbb{N}}$ embeds homeomorphically.
- The index set of the set of all metric spaces into which $2^{\mathbb{N}}$ embeds isometrically.
- The index set of the set of all metric spaces into which $\mathbb{N}^{\mathbb{N}}$ embeds homeomorphically.
- The index set of the set of all metric spaces into which $\mathbb{N}^{\mathbb{N}}$ embeds isometrically.

Thewmorakot conjectured a positive answer to the following.

Question 4.37 If X is an infinite computably presentable metric space, does it follow that the index set of $\{M \mid X \hookrightarrow M\}$ is Σ^1_1-complete?

The following results describing the complexity of various kinds of metric spaces are due to Melnikov and Nies [12].

Proposition 4.38

1. *The index set of the class of all infinite metric spaces is Π^0_2-complete.*
2. *The index set of all compact metric spaces is Π^0_3-complete.*
3. *The index set of connected metric spaces is Π^1_1-hard.*
4. *The index set of compact and connected metric spaces is Π^0_3-complete.*

It is rather straightforward to see that all of these sets have the appropriate level of complexity; it is only the completeness or the hardness that requires more than observation.

The first of these can be seen by defining a reduction of the standard index set Tot to it. The second can be proven by fixing a Π^0_3 subset S and constructing a uniformly c.e. family $(C_i)_{i\in\mathbb{N}}$ of substructures of Baire space so that $i \in S$ if and only if the metric space associated to C_i is compact. The proof of the next statement requires one to encode the problems in terms of whether a computable tree in Baire space has an infinite branch, which is known to be Π^1_1-complete. The last result requires one to code subspaces of the Hilbert cube that are connected and compact into a Π^0_3-set.

Finally, we mention the following result of Nies and Solecki [17].

Theorem 4.39 *The index set of all locally compact metric spaces is* Π^1_1*-complete.*

References

1. Alexander G. Melnikov. Computably isometric spaces. *J. Symbolic Logic*, 78(4):1055–1085, 2013.
2. Teerawat Thewmorakot. *Computability Theory on Polish Metric Spaces*. Ph.D. thesis, University of Connecticut, Storrs, CT, 2023.
3. Julien Melleray. Some geometric and dynamical properties of the Urysohn space. *Topology Appl.*, 155(14):1531–1560, 2008.
4. Vladimir Uspenskij. The Urysohn universal metric space is homeomorphic to a Hilbert space. *Topology Appl.*, 139(1-3):145–149, 2004.
5. Bradd Hart. An introduction to continuous model theory. *Model Theory of Operator Algebras*, 11:83, 2023.
6. Alexander Usvyatsov. Generic separable metric structures. *Topology and its Applications*, 155(14):1607–1617, 2008.
7. Paul Urysohn. Sur un espace métrique universel. *Bull. Sci. Math.*, pages 43–64, 1927. Prepared by Paul Alexandroff.
8. Paul Urysohn. Sur un espace métrique universel. *Bull. Sci. Math.*, pages 74–90, 1927. Prepared by Paul Alexandroff.
9. Miroslav Hušek. Urysohn universal space, its development and Hausdorff's approach. *Topology Appl.*, 155(14):1493–1501, 2008.
10. Itaï Ben Yaacov. Fraïssé limits of metric structures. *The Journal of Symbolic Logic*, 80(1):100–115, 2015.
11. Alexander G. Melnikov and Keng Meng Ng. Computable structures and operations on the space of continuous functions. *Fund. Math.*, 233(2):101–141, 2016.
12. Alexander G. Melnikov and André Nies. The classification problem for compact computable metric spaces. In *The Nature of Computation*, volume 7921 of *Lecture Notes in Comput. Sci.*, pages 320–328. Springer, Heidelberg, 2013.
13. Rodney G. Downey and Alexander G. Melnikov. Computably compact metric spaces. *Bull. Symb. Log.*, 29(2):170–263, 2023.
14. Wesley Calvert. The isomorphism problem for classes of computable fields. *Arch. Math. Logic*, 43(3):327–336, 2004.
15. Denis R. Hirschfeldt, Bakhadyr Khoussainov, Richard A. Shore, and Arkadii M. Slinko. Degree spectra and computable dimensions in algebraic structures. *Ann. Pure Appl. Logic*, 115(1-3):71–113, 2002.

16. Gerald E. Sacks. *Higher Recursion Theory*. Perspectives in Mathematical Logic. Springer-Verlag, Berlin, 1990.
17. André Nies and Slawomir Solecki. Local compactness for computable Polish metric spaces is Π^1_1-complete. In *Evolving Computability*, volume 9136 of *Lecture Notes in Comput. Sci.*, pages 286–290. Springer, Cham, 2015.

Chapter 5
Banach Spaces: $C(X)$-Spaces

In this chapter, we study the Banach spaces of the form $C(X; \mathbb{F})$ for X a compact metrizable Polish space and $\mathbb{F} \in \{\mathbb{R}, \mathbb{C}\}$. Recall that this is the space of all continuous functions from X into $\mathbb{F}$ in which the operations are defined pointwise and the norm is the supremum norm. We call these spaces *C-spaces*.

Again, we organize our presentation around the themes of computable presentability, computable categoricity, and index sets. Before proceeding with these topics, we review some useful classical material on these spaces. We also discuss some work on reducing the signature under consideration.

Throughout this section, X and Y will denote compact metric spaces.

5.1 Classical Results

There are two classical results related to isomorphisms of C-spaces that will be useful to us here; proofs of these results can be found in [1]. The first result is the Banach-Stone Theorem.

Theorem 5.1 (Banach-Stone Theorem) *Given a function $T : C(X; \mathbb{F}) \to C(Y; \mathbb{F})$, we have that T is an isomorphism if and only if there is a homeomorphism ϕ of Y onto X and a unimodular scalar $\lambda \in \mathbb{F}$ so that $T(f) = \lambda(f \circ \phi)$ for all $f \in C(X; \mathbb{F})$.*

Thus, an isomorphism of C-spaces is a weighted composition operator. This principle is especially useful when studying $C([0, 1]; \mathbb{F})$ since the homeomorphisms of $[0, 1]$ onto itself are simply the strictly monotonic functions from $[0, 1]$ onto $[0, 1]$.

The second result is the Stone-Weierstrass Theorem, which we will use to build presentations of C-spaces. Recall that a set $A \subseteq C(X; \mathbb{F})$ *separates points* if whenever $p, q \in X$ are distinct, there exists $f \in A$ so that $f(p) \neq f(q)$. This theorem will also be very useful to us in our treatment of commutative operator algebras in Chap. 7.

J. N. Y. Franklin et al., *Effective Metric Structure Theory*, Springer Monographs in Mathematics, https://doi.org/10.1007/978-3-032-13322-9_5

Theorem 5.2 (Stone-Weierstrass Theorem) *Suppose $A \subseteq C(X; \mathbb{F})$ separates points and contains the constant function 1. Then the algebra generated by A (that is, the smallest subset of $C(X; \mathbb{F})$ that includes A and is closed under multiplication, addition, scalar multiplication, and conjugation) is dense in $C(X; \mathbb{F})$.*

5.2 Computable Presentability

The following is a direct result of Theorem 7.11. Although this theorem appears in a later chapter, no circularity of reasoning results from using it here.

Proposition 5.3 *If X is a metric space with a computably compact presentation, then $C(X; \mathbb{F})$ is computably presentable.*

In light of the Banach-Stone Theorem, it seems reasonable to ask if $C(X; \mathbb{F})$ is computably presentable when X is homeomorphic to a metric space with a computably compact presentation. Recently, Melnikov and Ng have announced a negative answer (see [2]).

It is easy to see that the converse of Proposition 5.3 is false. Namely, take X to be the metric space consisting of two points in the plane at an incomputable distance from each other. Then X has no computable presentation, but $C(X; \mathbb{F})$ clearly does. Recently, Bazhenov et al. showed that for Stone spaces, the converse does hold [3].

Theorem 5.4 *Suppose Z is a Stone space. Then, $C(Z; \mathbb{F})$ is computably presentable if and only if Z is computably presentable.*

Proof As per the main result of [3], it suffices to show that the Boolean algebra of clopen sets has a computable presentation. We then turn to a classical result of Downey and Jockusch that every Δ^0_2-presentable Boolean algebra in which the atoms are Δ^0_2 is computably presentable [4]. One now uses a computable presentation of $C(X; \mathbb{F})$ and the total disconnectedness of Z to build a Δ^0_2 search procedure for these atoms. □

We are thus left with the following:

Problem 5.5 Classify the compact metrizable spaces X for which $C(X; \mathbb{F})$ is computably presentable.

We now turn to the computable presentability of some specific C-spaces. First, there are several natural computable presentations of the Banach space $C([0, 1]; \mathbb{F})$. For instance, as in Example 3.18, we may take the special points to be the power functions $x \mapsto x^n$ for $n \in \mathbb{N}$. In this case, the rational points are the rational polynomials; these are seen to be dense by the Stone-Weierstrass Theorem, and the computabilty of the supremum norm on these functions is a fairly easy exercise. A less natural, but in many ways easier to use, presentation is given by the standard Faber-Schauder basis for $C([0, 1]; \mathbb{F})$. The 0th distinguished point of this presentation is the constant function $x \mapsto 1$. If $n = 2^k + j$ for k a positive integer and $j \in \{1, \ldots, 2^k\}$, then the nth distinguished point of this presentation is

$$x \mapsto 2^{n-1} \int_0^x \chi_{[(2j-2)/2^{k+1},(2j-1)/2^{k+1}]}(t) - \chi_{[(2j-1)/2^{k+1},2j/2^{k+1}]}(t)\, dt.$$

So, for example, the 4th distinguished point of this presentation is a tent function with peak at $(3/4, 1)$ and base $[1/2, 1]$. Each vector in $C([0, 1]; \mathbb{F})$ can be uniquely expressed as an infinite linear combination of these functions; see, for example, [5]. Computability of the norm for this presentation is easy to demonstrate. Finally, we may present $C([0, 1]; \mathbb{F})$ by taking the special points to be the rational polygonal functions on $[0, 1]$. In this case, the rational points and the special points coincide, and again it is easy to see that the norm is computable on the rational points. These presentations are easily seen to be computably isomorphic via the identity map. Thus, we may take any one of them to be the standard presentation of $C([0, 1]; \mathbb{F})$.

The other particular Banach space we will study is $C(2^{\mathbb{N}}; \mathbb{F})$. In this case, we define the standard presentation by taking the special points to be the indicator functions of the cylinders. In other words, the rational points are the rational simple functions.

5.3 Computable Categoricity and Relative Categoricity

Nothing is known about computable categoricity of C-spaces in general. We focus here on two particular spaces: $C([0, 1]; \mathbb{F})$ and $C(2^{\mathbb{N}}; \mathbb{R})$. We also discuss relevant results for the Banach *algebra* $C([0, 1]; \mathbb{F})$.

5.3.1 $C([0, 1]; \mathbb{F})$

Incomputable Categoricity

In Chap. 4, we discussed Melnikov's 2013 result that $C([0, 1]; \mathbb{R})$ is not computably categorical as a metric space. Three years later, Melnikov and Ng proved that it is not computably categorical as a Banach space either. This is a direct consequence of the following [6].

Theorem 5.6 *There is a computable presentation of the Banach space $C([0, 1]; \mathbb{F})$ for which the unit is not a computable point.*

Recall that the unit of $C(X; \mathbb{F})$ is the constant function $t \mapsto 1$.

Proof We begin by fixing a computable pairwise disjoint sequence $(J_m)_{m\in\mathbb{N}}$ of closed subintervals of $[0, 1]$. We then construct a linearly dense sequence $(\sigma_n)_{n\in\mathbb{N}}$ of rational polygonal functions so that σ_n is constant on J_m when $m > n$.

Although the σ_n's give us a computable presentation of $C([0, 1]; \mathbb{F})$, this is not the presentation we seek. In fact, this presentation is computably isomorphic to the standard presentation. As in the proofs of Theorems 4.13 and 4.17, the special points of our presentation are constructed dynamically. That is, we construct a computable array $(\tau_{n,s})_{n,s\in\mathbb{N}}$ of rational polygonal functions that satisfies the following conditions:

1. $\tau_{n,0} = \sigma_n$.
2. $\lim_s \tau_{n,s}$ exists for each s.
3. When $\alpha_0, \ldots, \alpha_k$ are rational scalars, there exists $s \in \mathbb{N}$ so that $\left\|\sum_j \alpha_j \tau_{j,s}\right\| = \left\|\sum_j \alpha_j \tau_j\right\|$. Furthermore, s can be computed from $\alpha_0, \ldots, \alpha_k$.
4. $(\tau_n)_{n\in\mathbb{N}}$ is linearly dense.

We then define $C([0, 1]; \mathbb{F})^\#$ by declaring its nth special point to be τ_n. The third condition above ensures that this presentation is computable.

Let $\mathbf{1}$ denote the unit of $C([0, 1]; \mathbb{F})$. Our diagonalization strategy to ensure that $\mathbf{1}$ is not a computable point of $C([0, 1], \mathbb{F})^\#$ is as follows. Fix $e \in \mathbb{N}$. We wait for a stage s at which the eth Turing machine produces, on input $e + 1$, a (code of a) sequence $\alpha_0, \ldots, \alpha_k$ so that $\left\|1 - \sum_j \alpha_j \tau_{j,s}\right\| < 2^{-(e+1)}$. We then choose $m > k$ to be larger than any number considered by this diagonalization procedure earlier in the construction. Thus, $\sum_j \alpha_j \tau_{j,s}$ is constant on J_m, and this constant is within $2^{-(e+1)}$ of 1. For $j \leq s$, we define $\tau_{j,s+1}$ so that it is like $\tau_{j,s}$ except we introduce a "dip" on J_m. We do this in such a way as to preserve all norms calculated so far as well as the linear density of our special points. □

By the Banach-Stone Theorem, the unit of $C(X; \mathbb{F})$ is preserved by automorphisms. Thus, since the unit is a computable point of the standard presentation of $C([0, 1]; \mathbb{F})$, we immediately obtain the following.

Corollary 5.7 *$C([0, 1]; \mathbb{F})$ is not computably categorical as a Banach space.*

From Proposition 4.12 and Theorem 5.6, we obtain the following.

Corollary 5.8 *There is a computable presentation of $C([0, 1]; \mathbb{R})$ as a Banach space for which multiplication is not a computable operation.*

Relative Computable Categoricity

Since $C([0, 1]; \mathbb{R})$ is not computably categorical, we now turn to the examination of the Turing degrees that are capable of computing an isomorphism of any computable presentation of $C([0, 1]; \mathbb{F})$ onto the standard presentation. Clearly, the degree of Kleene's $\mathcal{O}$ is such a degree. However, by building on prior work of Brown [7], Franklin et al. proved the following [8].

Theorem 5.9 *$C([0, 1]; \mathbb{R})$ is Δ^0_5-categorical.*

We now outline the proof of Theorem 5.9. The first step is the following result by Brown [7].

Theorem 5.10 *Suppose that $C([0, 1]; \mathbb{R})^\#$ is a computable presentation of $C([0, 1]; \mathbb{R})$, and let $X \subseteq \mathbb{N}$. Furthermore, assume an X-computable point of $C([0, 1]; \mathbb{R})^\#$, f, is a homeomorphism of $[0, 1]$ with itself. Then the composition operator $g \mapsto g \circ f$ is an $(X \oplus 0'')$-computable map from $C([0, 1]; \mathbb{R})$ to $C([0, 1]; \mathbb{R})^\#$.*

To prove Theorem 5.10, we need the following preliminary result from [7]. Let $\mathbf{1}$ denote the constant function $t \mapsto 1$ on $[0, 1]$.

Proposition 5.11 *Suppose $C([0, 1]; \mathbb{R})^\#$ is a computable presentation.*

1. *$\mathbf{1}$ is a $\mathbf{0}'$-computable point of $C([0, 1]; \mathbb{R})^\#$.*
2. *The modulus operation $f \mapsto |f|$ is a $\mathbf{0}''$-computable operation of $C([0, 1]; \mathbb{R})^\#$.*

Proof (Theorem 5.10) For the sake of this proof, we take the Faber-Schauder basis functions to be the special points of $C([0, 1]; \mathbb{R})$ (that is, the standard presentation of $C([0, 1]; \mathbb{R})$). Let $(F_n)_{n\in\mathbb{N}}$ be a standard one-to-one enumeration of this basis. Without loss of generality, assume $F_0 = \mathbf{1}$. Suppose f is a homeomorphism of $[0, 1]$ with itself, and assume also that f is an X-computable point of $C([0, 1]; \mathbb{R})^\#$. Since the operator norm of $g \mapsto g \circ f$ is 1, it suffices to show that $F_n \circ f$ is an $X \oplus 0''$-computable point of $C([0, 1]; \mathbb{R})^\#$ uniformly in n. To this end, we only need to show that if T is one of the tent functions that appears as part of a Faber-Schauder basis function, then $T \circ f$ is an $X \oplus 0''$-computable point of $C([0, 1]; \mathbb{R})^\#$ uniformly in T. Consequently, we assume T has the form

$$T(x) = \begin{cases} 0 & x < p \\ \frac{2}{q-p}(x-p) & p < x < \frac{p+q}{2} \\ \frac{-2}{q-p}(x-q) & \frac{p+q}{2} \le x < q \\ 0 & q \le x \end{cases}$$

where p, q are dyadic rationals. It is easy to see (quite literally if a picture is drawn) that

$$T(x) = \min\left\{\max\left\{\frac{2}{q-p}(x-p), 0\right\}, \max\left\{\frac{-2}{q-p}(x-q), 0\right\}\right\}.$$

Thus, when $g \in C([0, 1]; \mathbb{R})$,

$$T \circ f = \min\left\{\max\left\{\frac{2}{q-p}(f - p\cdot\mathbf{1}), \mathbf{0}\right\}, \max\left\{\frac{-2}{q-p}(f - q\cdot\mathbf{1}), \mathbf{0}\right\}\right\}.$$

Since, by Proposition 5.11, the modulus operation is a $\mathbf{0}''$-computable operation of $C([0, 1]; \mathbb{R})^\#$, it follows that max and min are $\mathbf{0}''$-computable operations of $C([0, 1]; \mathbb{R})^\#$. Also by Proposition 5.11, $\mathbf{1}$ is a $\mathbf{0}''$-computable point of $C([0, 1]; \mathbb{R})^\#$. The desired conclusion follows. □

In light of Theorem 5.10, we are now led to determine the possible complexity of a homeomorphism of [0, 1] as a point of a computable presentation $C([0, 1]; \mathbb{R})$. Our first step in this direction is the following. Let $\mathfrak{M}$ denote the set of all monotone functions on [0, 1].

Lemma 5.12 *$\mathfrak{M}$ is a Π^0_4 closed set of every computable presentation of $C([0, 1]; \mathbb{R})$.*

Proof Let $C([0, 1]; \mathbb{R})^{\#}$ be a computable presentation. We first note that an open ball of $C([0, 1]; \mathbb{R})$ contains a monotone function if and only if it contains a strictly monotone function. Thus, it suffices to show that the set of all rational open balls of $C([0, 1]; \mathbb{R})^{\#}$ that contain a strictly monotone function is Π^0_4. To this end, for each nonconstant $g \in C([0, 1]; \mathbb{R})$, we let $\widehat{g} = (\max(\text{ran}(g)) - \min(\text{ran}(g)))(g - \min(\text{ran}(g)))$. Thus, $\text{ran}(\widehat{g}) = [0, 1]$, and it follows that g is strictly monotone if and only if $\widehat{g}$ is a homeomorphism of [0, 1] with itself. We then note that the composition map $f \mapsto f \circ \widehat{g}$ is surjective if and only if $\widehat{g}$ is injective. By putting these observations together, we see that a rational open ball B of $C([0, 1]; \mathbb{R})^{\#}$ contains a homeomorphism of [0, 1] with itself if and only if, for every finite collection of rational open balls $\{B_j \mid j < N\}$ of $C([0, 1]; \mathbb{R})^{\#}$, there is a nonconstant rational point $h \in B$ of $C([0, 1]; \mathbb{R})^{\#}$ such that, for each $j < N$, there is a rational point ρ of $C([0, 1]; \mathbb{R})$ so that $\rho \circ \widehat{h} \in B(g_j, \delta_j)$. Here, we use the Faber-Schauder basis for the standard presentation of $C([0, 1]; \mathbb{R})$. The proof of one direction of this claim has already been outlined. For details of the proof of the converse, we refer the reader to [8, Claim 5.12]. By the results of Brown above, this condition can be seen to be $\Pi^0_2(\mathbf{0}'')$. □

We are now set up to prove the following.

Lemma 5.13 *If $C([0, 1]; \mathbb{R})^{\#}$ is a computable presentation, then at least one of its Π^0_4-computable points is a homeomorphism of* [0, 1] *with itself.*

Proof By means of Lemma 5.12, it is at least possible to show $\mathfrak{M}$ contains a Π^0_4-computable point of $C([0, 1]; \mathbb{R})^{\#}$. Unfortunately, this point may not be injective. However, we can still achieve an injective limit if we carefully modify our search procedure as follows. Let $(U_s)_{s \in \mathbb{N}}$ be an effective enumeration of the rational open balls of $C([0, 1]; \mathbb{R})^{\#}$. By means of Lemma 5.12, it is possible to construct a Π^0_4 sequence $(B_s)_{s \in \mathbb{N}}$ of rational open balls that satisfies the following conditions:

1. $B_s \cap \mathfrak{M} \neq \emptyset$.
2. $B_{s+1} \subseteq B_s$.
3. $\text{diam}(B_s) \leq 2^{-s}$.
4. For each $s \in \mathbb{N}$, there is a rational vector ρ of $C[0, 1]$ (in the standard presentation) such that $\rho \circ \widehat{\phi} \in U_s$ for each $\phi \in B_s$.

Again, we use the Faber-Schauder basis vectors for our standard presentation of $C([0, 1]; \mathbb{R})$. The details of this construction can be found in [8]. By completeness, there is a unique $\phi \in \bigcap_s B_s$, and the last of these conditions ensures that $f \mapsto f \circ \widehat{\phi}$ is surjective. Thus, $\widehat{\phi}$ is a homeomorphism of [0, 1] with itself. □

Theorem 5.9 now follows from Theorem 5.10 and the previous two lemmas. Theorem 5.9 is the best result to date on the relative categoricity of $C([0, 1]; \mathbb{R})$. We are thus led to the following.

Question 5.14 Is the bound in Theorem 5.9 optimal? In particular, does $C([0, 1]; \mathbb{R})$ have a degree of categoricity, and if so, what is it?

Banach Algebra Results

We now consider $C([0, 1]; \mathbb{R})$ as a Banach algebra. Our standard presentation of this structure is the same as its standard presentation as a Banach space. Our main theorem regarding this structure is the following result of Melnikov and Ng [6].

Theorem 5.15 $C([0, 1]; \mathbb{R})$ *is not computably categorical as a Banach algebra.*

We proved $C([0, 1]; \mathbb{R})$ is not computably categorical as a Banach space by constructing a computable presentation of which the unit is not a computable point. Unfortunately, this strategy is not available to us here because of the following, which is essentially Theorem 4.1 of [9].

Theorem 5.16 *If $\mathcal{B}$ is a commutative unital Banach algebra, then the unit of $\mathcal{B}$ is a computable point of every computable presentation of $\mathcal{B}$.*

Proof Let $\mathbf{1}_{\mathcal{B}}$ denote the unit of $\mathcal{B}$, and let $\mathcal{B}^{\#}$ be a computable presentation. Fix a rational vector v_0 of $\mathcal{B}$ so that $\mathbf{1}_{\mathcal{B}}$ is sufficiently close to v_0. Then, to compute a rational vector that is within 2^{-k} of $\mathbf{1}_B$, we search for a rational vector v so that $\|v^2 - v\|$ is sufficiently small and v is not too far from v_0. Appropriate values for these error bounds can be found in the proof of [9, Theorem 4.1]. A key point in the verification is that, by the Gelfand Theorem, we can represent B as a C-space. □

Considering the role of the identity map suggests an alternative strategy to demonstrating Theorem 5.15. We first note that the identity map is a computable point of the standard presentation of $C([0, 1]; \mathbb{R})$. Thus, its image under any computable map is computable. We then note that, by the Banach-Stone Theorem, the image of the identity map under an isomorphism $T : C([0, 1]; \mathbb{R}) \to C([0, 1]; \mathbb{R})$ is a strictly monotone function. Thus, Theorem 5.15 is a corollary of the following (also proved in [6]).

Theorem 5.17 *There is a computable presentation $C([0, 1]; \mathbb{R})^{\dagger}$ of $C([0, 1]; \mathbb{R})$ as a real Banach algebra for which no strictly monotonic function is a computable point.*

Proof As the vertical-distortion strategy used in the proof of Theorem 5.6 is not available to us, we deploy a horizontal-distortion strategy. Namely, we define an increasing and bounded sequence $(\alpha_s)_{s\in\mathbb{N}}$ of rational numbers that are no smaller than 1, and we set $\alpha = \lim_s \alpha_s$. The real α is left c.e. but is not guaranteed to be computable.

We note that $C([0,\alpha];\mathbb{R})$ is isomorphic to $C([0,1];\mathbb{R})$ via the compression map $x \mapsto \alpha^{-1}x$. Thus, a computable presentation $C([0,\alpha];\mathbb{R})^{\#}$ induces a corresponding computable presentation $C([0,1];\mathbb{R})^{\dagger}$ so that these presentations are computably isomorphic. Since this isomorphism preserves monotonicity, it suffices to construct a presentation $C([0,\alpha];\mathbb{R})^{\#}$ so that none of its computable points are increasing.

Thus, for each $e \in \mathbb{N}$, let R_e be the requirement that e is not a $C[0,\alpha]^{\#}$-index of an increasing function. Set $\epsilon_e = 2^{-(e+1)}$. If a requirement R_e acts at a stage s, it will define α_{s+1} to be $\alpha_s + \epsilon_e$. Thus, our choice of ϵ_e's ensures that $(\alpha_s)_{s\in\mathbb{N}}$ is bounded.

As in the proof of Theorem 5.6, we define the special points of $C([0,\alpha];\mathbb{R})^{\#}$ dynamically, that is, we define an array $(h_{j,s})_{j,s\in\mathbb{N}}$ and take $h_j = \lim_s h_{j,s}$ to be the jth special point of $C([0,\alpha];\mathbb{R})^{\#}$.

We say that N_e *requires attention at stage* s if on a sufficiently large $k \leq s$, the eth Turing machine outputs a code of a sequence $(\beta_0,\dots,\beta_s)$ of rational scalars so that $\sum_{j\leq s} h_{j,s}(\beta_s)$ is sufficiently larger than $\sum_{j\leq s} h_{j,s}(\beta_s - e)$. Details on how large these quantities need to be can be found via a careful examination of the proof in [6].

At stage 0, we set $\alpha_0 = 1$ and $h_{j,0}$ to equal the jth term of the Faber-Schauder basis for $C([0,1];\mathbb{R})$. We assume $h_{0,0}$ is the unit.

At stage $s+1$, we search for the least $e \leq s$ such that N_e requires attention at stage s. If there is no such e, then we go on to the next stage without making any changes. Otherwise, we define α_{s+1} to be $\alpha_s + \epsilon_e$. Then, for each $j \leq s$ and $x \in [0,\alpha_{s+1}]$, we define

$$h_{j,s+1}(x) = \begin{cases} h_{j,s}(x) & x \leq \alpha_s \\ h_{j,s}(x-\epsilon_e) & \alpha_s < x \leq \alpha_{s+1} \end{cases}.$$

The point is that any function f such that $\left\| f - \sum_j \beta_j h_{j,s+1} \right\| < 2^{-k}$ will fail to be increasing on $[\alpha_s, \alpha_{s+1}]$. In addition, this action preserves all norms calculated at prior stages. This strategy also ensures that the algebra generated by the h_j's contains the constant functions and separates points. Hence, it indeed produces a presentation of $C([0,1];\mathbb{R})^{\#}$. □

5.3.2 $C(2^{\mathbb{N}};\mathbb{R})$

Now we consider the Banach space $C(2^{\mathbb{N}};\mathbb{R})$. As with the metric space $C(2^{\mathbb{N}})$, Thewmorakot obtained results parallel to Melnikov and Ng's [10].

Theorem 5.18 *There is a computable presentation of $C(2^{\mathbb{N}};\mathbb{R})$ as a real Banach space for which the unit is not a computable point.*

In this case, we are able to use as our standard structure the rational simple functions $(\ell_i)_{i\in\mathbb{N}}$ as in the proof of Theorem 4.20; this contrasts with the change in standard structures between Theorems 4.17 and 5.6. We also use the absolute value function $f \mapsto |f|$ as our auxiliary function to show that 1 is not computable; the approach to satisfying all the requirements is quite similar.

As a corollary, we have the following:

Corollary 5.19 $C(2^{\mathbb{N}}; \mathbb{R})$ *is not computably categorical as a real Banach space.*

So far, the story for $C(2^{\mathbb{N}})$ has paralleled the story for $C[0, 1]$. However, once we allow for multiplication, the story becomes quite different.

Theorem 5.20 $C(2^{\mathbb{N}}; \mathbb{R})$ *is computably categorical as a real Banach algebra.*

This theorem is a corollary of two theorems. The first is a theorem of McNicholl, which states that $C(2^{\mathbb{N}}; \mathbb{C})$ is computably categorical as a C*-algebra (see [11]). The second is a theorem of Fox [12, Theorem 4.5], which states that $C(X; \mathbb{C})$ is computably categorical as a C*-algebra if and only if $C(X; \mathbb{R})$ is computably categorical as a real Banach algebra.

5.4 Index Sets

Little is known about the index sets of C-spaces. At present, the only result of this kind is the following theorem of Franklin et al. [8].

Theorem 5.21 *The index set of* $C([0, 1]; \mathbb{R})$ *is* Σ^0_7.

Recall from Chap. 3 that the index set of $C([0, 1]; \mathbb{R})$ is the set of all indices of presentations of $C([0, 1]; \mathbb{R})$. Note that the index set of Banach space presentations is Π^0_2 (see [13]).

Theorem 5.21 is an immediate consequence of Theorem 5.9 by means of which quantification over all of Baire space may be reduced to quantification over an arithmetical set.

On one hand, Theorem 5.21 is remarkable in that Σ^0_7 is fairly low in the arithmetical hierarchy, and certainly much lower than the obvious upper bound of Σ^1_1. Nevertheless, we are still left with the following question.

Question 5.22 Is the upper bound of Σ^0_7 for the index set of $C([0, 1]; \mathbb{R})$ sharp? In other words, is this index set Σ^0_7-complete?

Since Theorem 5.21 is our only C-space index set result to date, we also pose the following.

Problem 5.23 Find the complexities of the index sets for other C-spaces such as $C(2^{\mathbb{N}}; \mathbb{F})$.

References

1. Richard J. Fleming and James E. Jamison. *Isometries on Banach Spaces: function spaces*, volume 129 of *Chapman & Hall/CRC Monographs and Surveys in Pure and Applied Mathematics*. Chapman & Hall/CRC, Boca Raton, FL, 2003.
2. Heer Tern Koh, Alexander G. Melnikov, and Keng Meng Ng. Counterexamples in effective topology. *The Journal of Symbolic Logic*. Published online 2025:1–24.
3. Nikolay Bazhenov, Matthew Harrison-Trainor, and Alexander Melnikov. Computable Stone spaces. *Ann. Pure Appl. Logic*, 174(9):Paper No. 103304, 25, 2023.
4. Rod Downey and Carl G. Jockusch. Every low Boolean algebra is isomorphic to a recursive one. *Proc. Amer. Math. Soc.*, 122(3):871–880, 1994.
5. N. L. Carothers. *A Short Course on Banach Space Theory*, volume 64 of *London Mathematical Society Student Texts*. Cambridge University Press, Cambridge, 2005.
6. Alexander G. Melnikov and Keng Meng Ng. Computable structures and operations on the space of continuous functions. *Fund. Math.*, 233(2):101–141, 2016.
7. Tyler A. Brown. Computing on the Banach space $C[0,1]$. *Computability*, 10(3):257–270, 2021.
8. Johanna N.Y. Franklin, Rupert Hölzl, Alexander Melnikov, Keng Meng Ng, and Daniel Turetsky. Computable classifications of continuous, transducer, and regular functions. *Theoret. Comput. Sci.*, 1032:Paper No. 115086, 28, 2025.
9. Peter Burton, Christopher J. Eagle, Alec Fox, Isaac Goldbring, Matthew Harrison-Trainor, Timothy H. McNicholl, Alexander Melnikov, and Teerawat Thewmorakot. Computable Gelfand duality. *Proc. Amer. Math. Soc.,* to appear.
10. Teerawat Thewmorakot. *Computability Theory on Polish Metric Spaces*. Ph.D. thesis, University of Connecticut, Storrs, CT, 2023.
11. Timothy H. McNicholl. Evaluative presentations. *J. Logic Comput.*, 35(5):Paper No. exaf036, 11, 2025.
12. Alec Fox. Computable presentations of C*-algebras. *J. Symb. Log.*, 89(3):1313–1338, 2024.
13. Tyler A. Brown, Timothy H. McNicholl, and Alexander G. Melnikov. On the complexity of classifying Lebesgue spaces. *J. Symb. Log.*, 85(3):1254–1288, 2020.

Chapter 6
Banach Spaces: L^p Spaces

In this chapter, we discuss computable presentations of another kind of Banach space, the Lebesgue spaces. Recall that a Banach space is an L^p *space* if it is isomorphic to $L^p(\Omega)$ for some measure space Ω and that a Banach space is a Lebesgue space if it is an L^p space for some $p \in [1, \infty]$. Our field of scalars $\mathbb{F}$ can be either $\mathbb{R}$ or $\mathbb{C}$. As is customary in measure theory, we identify sets that differ only on a set of measure zero, and we identify functions that differ only on a set of measure zero.

6.1 Classical Theory of L^p Spaces

6.1.1 Classifying Separable L^p Spaces

One of our first goals (see Sect. 6.2) is to classify the computably presentable Lebesgue spaces. Since every computably presentable structure is separable, it makes sense to first understand which Lebesgue spaces are separable. Fortunately, there is a complete classification of the separable Lebesgue spaces, which we endeavor to describe now.

Our first step towards stating this classification is to note that L^∞ spaces are separable only when they are finite dimensional. So, henceforth, we will direct our attention to values of p in $[1, \infty)$.

Our next step towards stating this classification is to define the L^p sums of Banach spaces for $1 \leq p < \infty$. To this end, suppose $\mathcal{B}_0$ and $\mathcal{B}_1$ are Banach spaces. We then let $\mathcal{B}_0 \oplus_p \mathcal{B}_1$ denote the Banach space whose underlying vector space is $\mathcal{B}_0 \times \mathcal{B}_1$ and whose norm is given by $\|(u, v)\|^p = \|u\|^p + \|v\|^p$. We call $\mathcal{B}_0 \times \mathcal{B}_1$ the L^p*-sum* of $\mathcal{B}_0$ and $\mathcal{B}_1$. If Ω_0 and Ω_1 are measure spaces, then it is not difficult to see that $L^p(\Omega_0) \oplus_p L^p(\Omega_1)$ is isomorphic to $L^p(\Omega_0 \oplus \Omega_1)$.

Recall that when $\Omega = (X, \Sigma, \mu)$ is a measure space, there is a pseudometric D on the measurable sets of Ω with finite measure given by $D(A, B) = D(A \triangle B)$. Recall

J. N. Y. Franklin et al., *Effective Metric Structure Theory*, Springer Monographs in Mathematics, https://doi.org/10.1007/978-3-032-13322-9_6

also that Ω is separable if the corresponding metric space is separable. Our third step towards classifying the separable Lebesgue spaces is to show that the separability of $L^p(\Omega)$ entails the separability of Ω. Although this result is surely well known, we were unable to find it in the literature and so we offer a proof below.

Proposition 6.1 *Suppose $1 \leq p < \infty$. If $L^p(\Omega)$ is separable, then Ω is separable.*

Proof To prove the proposition, write $\Omega = (X, \Sigma, \mu)$. Suppose $\mathcal{F}$ is a countable dense set of vectors in $L^p(\Omega)$. For each $f \in \mathcal{F}$ and each $k \in \mathbb{N}$, set

$$D_{f,k} = \{t \in X \ : \ |f(t)| \geq 2^{-k}\}.$$

Thus, $\mu(D_{f,k}) < \infty$. Let $\mathcal{D} = \{D_{f,k} \ : \ f \in \mathcal{F} \ \wedge \ k \in \mathbb{N}\}$.

We claim that $\mathcal{D}$ is dense in Ω. To see this, suppose E is measurable and $\mu(E) < \infty$. Thus, $\chi_E \in L^p(\Omega)$. Since $\mathcal{F}$ is dense in $L^p(\Omega)$, there is a sequence $(f_n)_{n \in \mathbb{N}}$ of vectors in $\mathcal{F}$ that converges to χ_E in the L^p-norm. Thus, $(f_n)_{n \in \mathbb{N}}$ converges to χ_E in measure.

Now let $0 < \epsilon < 1$. Choose $k, n \in \mathbb{N}$ so that $2^{-k} + (1 - 2^{-k})^{-p} 2^{-kp} < \epsilon$ and so that $\|f_n - \chi_E\|_p < 2^{-k}$. Furthermore, we choose k, n large enough that

$$\mu(\{t \in X \ : \ |f_n(t) - \chi_E(t)| \geq 2^{-k}\}) < 2^{-k}.$$

We claim that $\mu(D_{f_n,k} \triangle E) < \epsilon$. We see that

$$D_{f_n,k} - E \subseteq \{t \in X \ : \ |f_n(t) - \chi_E(t)| \geq 2^{-k}\}.$$

Then $\mu(D_{f_n,k} - E) < 2^{-k}$. If $t \in E - D_{f_n,k}$, then $|f_n(t) - \chi_E(t)| \geq 1 - 2^{-k}$. By Chebychev's Inequality,

$$\mu(E - D_{f_n,k}) \leq (1 - 2^{-k})^{-p} 2^{-kp}.$$

Thus, $\mu(D_{f_n,k} \triangle E) < 2^{-k} + (1 - 2^{-k})^{-p} 2^{-kp} < \epsilon$, as desired. □

It is well known that every nonzero separable L^2 space is isomorphic either to ℓ^2 or ℓ^2_n for some n. By means of Proposition 6.1 and the Carathéodory classification of separable measure spaces, we now obtain the following classification of L^p spaces for $p \neq 2$ (more details can be found in [1]).

Theorem 6.2 (Classification of separable L^p spaces, $p \neq 2$) *Suppose $1 \leq p < \infty$ and $p \neq 2$. Then every nonzero separable L^p space is isomorphic to exactly one of the following.*

1. ℓ^p_n *for some $n \geq 1$. In this case, the underlying measure space is purely atomic and has exactly n atoms.*
2. ℓ^p. *In this case, the underlying measure space is purely atomic and has $\aleph_0$ atoms.*
3. $L^p[0, 1]$. *In this case, the underlying measure space is nonatomic.*

4. $\ell_n^p \oplus_p L^p[0, 1]$ *for some* $n \geq 1$. *In this case, the underlying measure space has exactly* n *atoms but is not purely atomic.*
5. $\ell^p \oplus_p L^p[0, 1]$. *In this case, the underlying measure space has* $\aleph_0$ *atoms but is not purely atomic.*

6.1.2 Disjointness of Support

We now discuss some classical material that will assist in computing vectors in a Lebesgue space with disjoint support. Our starting point is the following.

Theorem 6.3 *Suppose* $1 \leq p < \infty$ *and* $p \neq 2$. *If* f, g *are vectors in an* L^p *space, then* f *and* g *are disjointly supported if and only if*

$$\|f + g\|_p^p + \|f - g\|_p^p = 2(\|f\|_p^p + \|g\|_p^p).$$

Theorem 6.3 was first proven by Hanner (though he credits this result to A. Beurling) [2]. It was later proven independently by Lamperti [3]. Hanner's proof reduces the problem to the calculation of a minimum by means of the techniques of multivariable calculus and some basic facts about convex functions. Lamperti's proof utilizes more sophisticated tools from the theory of convex functions such as Young's Inequality.

An immediate consequence of Theorem 6.3 is the following.

Theorem 6.4 *If* $1 \leq p < \infty$ *and* $p \neq 2$, *then every linear isometric map from an* L^p *space to an* L^p *space preserves disjointness of support.*

For the sake of computing disjointly supported vectors from a presentation of a Lebesgue space, we seek to develop a stable version of Theorem 6.3. By this we mean that, informally speaking, if f, g are close to being disjointly supported (that is, if $\|f + g\|_p^p + \|f - g\|_p^p = 2(\|f\|_p^p + \|g\|_p^p)|$ is sufficiently small), then f, g are close to vectors f', g' that are disjointly supported. To this end, for $w, z \in \mathbb{C}$, set

$$\sigma_{1,p}(z, w) = |2(|z|^p + |w|^p) - (|z - w|^p + |z + w|^p)|.$$

When $p \neq 2$, set $c_p = |4 - 2\sqrt{2}^p|^{-1}$, and set $\sigma_p(z, w) = c_p \sigma_{1,p}(z, w)$. The following is proven in [4].

Theorem 6.5 *Suppose* $p \geq 1$ *and* $p \neq 2$. *Then*

$$\min\{|z|^p, |w|^p\} \leq \sigma_p(z, w) \tag{6.1}$$

for all $w, z \in \mathbb{C}$. *Furthermore, if* $1 \leq p < 2$, *then*

$$2|z|^p + 2|w|^p - |z + w|^p - |z - w|^p \geq 0,$$

and if $2 < p$, *then*

$$2|z|^p + 2|w|^p - |z + w|^p - |z - w|^p \leq 0.$$

Theorem 6.3 follows immediately from Theorem 6.5, and the proof of Theorem 6.5 is similar in spirit to Hanner's proof of Theorem 6.3. Theorem 6.5 also leads to the following, which is our stable version of Theorem 6.3. A proof is given in [4].

Theorem 6.6 *Suppose* $1 \leq p < \infty$ *and* $p \neq 2$. *Let* f, g *be vectors of an* L^p *space* $\mathcal{B}$. *Then there exist disjointly supported* $f_0, g_0 \in \mathcal{B}$ *satisfying*

$$\max\{\|f - f_0\|^p, \|g - g_0\|^p\} \leq \sigma_p(\|f\|, \|g\|).$$

6.1.3 Classification of Linear Isometries

Lamperti used Theorem 6.4 to prove his classification of the isomorphisms of L^p spaces with $p \neq 2$ [3]. This classification will be of use to us when discussing computable categoricity of L^p spaces, so we will spend some time describing it now. We begin with some preliminary measure-theoretic terminology. When Ω is a measurable space, let $\mathcal{F}(\Omega)$ denote the set of all $\mathbb{F}$-valued measurable functions on the points of Ω.

Definition 6.7 Suppose $\Omega_j = (X_j, \Sigma_j, \mu_j)$ is a measure space for $j \in \{1, 2\}$. A function $T : \Sigma_1 \to \Sigma_2$ is a *regular set isomorphism* if it satisfies the following conditions modulo sets of measure 0.

- $T(X_1 \setminus A) = X_2 \setminus T(A)$.
- $T(\bigcup_n A_n) = \bigcup_n T(A_n)$ if $A_0, A_1, \ldots$ are pairwise disjoint.
- $\mu_2(T(A)) = 0$ if and only if $\mu_1(A) = 0$.

Regular set isomorphisms naturally extend to mappings of measurable scalar-valued functions as follows.

Theorem 6.8 *Suppose* $\Omega_j = (X_j, \Sigma_j, \mu_j)$ *is a measure space for* $j \in \{1, 2\}$, *and suppose* $T : \Sigma_1 \to \Sigma_2$ *is a regular set isomorphism. Then there is a unique map* T_+ *from* $\mathcal{F}(\Omega_1)$ *into* $\mathcal{F}(\Omega_2)$ *so that, modulo sets of measure* 0, $T_+(f)^{-1}[B] \cap T(X_1) = T(f^{-1}[B])$ *whenever* B *is a Borel subset of* $\mathbb{F}$ *and so that* $T_+(f)(p) = 0$ *when* $p \in X_2 \setminus T(X_1)$.

The map T_+ will be called the *natural extension* of T.

Now we use Theorem 6.8 to develop a characterization of linear isometries between L^p spaces.

Theorem 6.9 *Suppose $\Omega_j = (X_j, \Sigma_j, \mu_j)$ is a σ-finite measure space for $j \in \{1, 2\}$. Suppose $0 < p < \infty$ and $p \neq 2$. Let $\Phi : L^p(\Omega_1) \to L^p(\Omega_2)$. Then Φ is a linear isometry if and only if there is a regular set isomorphism $T : \Sigma_1 \to \Sigma_2$ and a measurable function $h : X_2 \to \mathbb{F}$ so that $\Phi = h \circ T_+$ and so that*

$$\mu_1(A) = \int_{T(A)} |h|^p d\mu_2$$

for every $A \in \Sigma_1$.

By considering the counting measure on $\mathbb{N}$, we obtain the following, which was first proven for real ℓ^p spaces by Banach [5].

Corollary 6.10 *Suppose $1 \leq p < \infty$ and $p \neq 2$. Then $\Psi : \ell^p \to \ell^p$ is a linear isometry if and only if there is a permutation ϕ of $\mathbb{N}$ and a sequence $(\lambda_n)_{n \in \mathbb{N}}$ of unimodular scalars so that $\Psi(f) = \sum_{n=0}^{\infty} \lambda_n f(n) e_{\phi(n)}$.*

6.1.4 Embeddings

The results on computable presentability of L^p spaces depend heavily on the embedding results in this section. Herein, in keeping with Chap. 2, all embeddings are linear and isometric.

Theorem 6.11 *Suppose $2 < p_1 < p_2 < \infty$.*

1. *No infinite-dimensional L^{p_2} space embeds into an L^{p_1} space.*
2. *No infinite-dimensional L^{p_1} space embeds into an L^{p_2} space.*

Part (1) of Theorem 6.11 was proven by Banach [5], and Part (2) of Theorem 6.11 was proven by Paley [6]. The next theorem was proven by Krivine in 1965 [7].

Theorem 6.12 *Suppose $1 \leq p < \infty$. Let $\mathcal{B}$ be a Banach space, and let r, s be positive integers so that $2(r-1) < p < 2r \leq 4s$. Then $\mathcal{B}$ embeds into an L^p space if and only if for all $v_1, \ldots, v_n \in \mathcal{B}$ and all $\alpha_1, \ldots, \alpha_n \in \mathbb{R}$,*

$$(-1)^r \sum_{\sigma \in \{-1,1\}^{2s}} \sum_{\tau \in \{1,\ldots,n\}^{2s}} \left\| \sum_{j<2s} \sigma(j) v_{\tau(j)} \right\|_{\mathcal{B}}^q \cdot \prod_{j<2s} \alpha_{\tau(j)} \geq 0$$

provided $\sum_j \alpha_j = 1$.

The following result contrasts with Theorem 6.11 and is due to Bretagnolle et al. [8].

Theorem 6.13 *If $1 \leq p < r < 2$, then $L^r[0,1]$ embeds into $L^p[0,1]$.*

Theorem 6.13 is proven by first showing that ℓ^r embeds into $L^p[0, 1]$. This is accomplished via a very elegant application of the theory of r-stable random variables. It is then shown by means of ultraproducts that $L^r[0, 1]$ isometrically embeds into $L^p[0, 1]$ when $1 \leq p < r < 2$. This was the first application of ultraproducts in functional analysis.

The following now proceeds from the classification of separable L^p spaces.

Corollary 6.14 *If $1 \leq p < r < 2$, then every separable L^r space embeds into $L^p[0, 1]$.*

6.1.5 Disintegrations

The theory of disintegrations of Lebesgue spaces was developed entirely for computability-theoretic purposes in [4, 9, 10]. However, the content of this theory is purely classical, and so we cover it here.

Suppose $f, g \in L^p(\Omega)$. Then f is a *subvector* of g if there is a measurable set A so that $f = g \cdot \chi_A$. We write $f \preceq g$ in this case. The concept of disintegration is then defined as follows. Recall that $A^* = \bigcup_{n \in \mathbb{N}} A^n$.

Definition 6.15 Suppose $1 \leq p < \infty$ and suppose $\mathcal{B}$ is an L^p-space. A *disintegration of $\mathcal{B}$* is a map $\phi : S \to \mathcal{B}$ that satisfies the following conditions:

1. S is a subtree of $\mathbb{N}^*$; that is, S is a nonempty subset of $\mathbb{N}^*$ that is closed under prefixes.
2. For all $\sigma, \tau \in S$, if $\sigma \subseteq \tau$, then $\phi(\tau) \preceq \phi(\sigma)$.
3. For all $\sigma, \tau \in S$, if σ and τ are incomparable, then $\phi(\sigma)$ and $\phi(\tau)$ are disjointly supported.
4. For each nonterminal node σ of S, $\phi(\sigma) = \sum_\tau \phi(\tau)$ where τ ranges over the children of σ in S; in other words, ϕ is *summative*.
5. The span of the range of ϕ is dense in $\mathcal{B}$.

Since the simple functions are dense in $L^p(\Omega)$, it is fairly easy to show that every L^p space has a disintegration. The key difficulty is to show that computable presentations of L^p spaces have computable disintegrations (see Theorem 6.39 below).

With disintegrations we have associated classes of isomorphisms and almost norm-maximizing chains that are used in the construction of linear isometries. These are defined as follows.

Definition 6.16 Suppose ϕ_j is a disintegration of $\mathcal{B}_j$ for $j \in \{0, 1\}$, and let $S_j = \mathrm{dom}(\phi_j)$. A map $f : S_0 \to S_1$ is an *isomorphism* of ϕ_0 and ϕ_1 if it is an order isomorphism (with respect to $\subseteq$) from S_0 onto S_1 and $\|\phi_1(f(\nu))\|_{\mathcal{B}_1} = \|\phi_0(\nu)\|_{\mathcal{B}_0}$ for all $\nu \in S_0$.

Definition 6.17 Suppose ϕ is a disintegration of an L^p space, and let $S = \mathrm{dom}(\phi)$. A chain $C \subseteq S$ is *almost norm-maximizing* if, whenever $\nu \in C$ is a nonterminal node of S, C contains a child ν' of ν so that

$$\max_{\mu} \|\phi(\mu)\|_p^p \leq \|\phi(\nu')\|_p^p + 2^{-|\nu|},$$

where μ ranges over the children of ν in S.

Since every disintegration ϕ is summative, it is not too hard to show that almost norm-maximizing chains exist. In fact, norm-maximizing chains exist. Again, the challenge is showing that computable presentations have computable norm-maximizing chains. It may not be possible to find a child that yields the largest possible norm among all children of a node in a disintegration. However, it turns out that one can effectively find a child whose norm is almost maximal.

The next proposition is from [9].

Proposition 6.18 *If $g_0 \succeq g_1 \succeq \ldots$ are vectors in an L^p space, then $\lim_n g_n$ exists in the L^p-norm and is the $\preceq$-infimum of $\{g_0, g_1, \ldots\}$.*

The following theorem was first proven for ℓ^p spaces in [4] and generalized to arbitrary L^p spaces in [10].

Theorem 6.19 *Suppose Ω is a measure space and $\phi : S \to L^p(\Omega)$ is a disintegration.*

1. *If $C \subseteq S$ is an almost norm-maximizing chain, then the $\preceq$-infimum of $\phi[C]$ exists and is either $\mathbf{0}$ or an atom of $\preceq$. Furthermore, the $\preceq$-infimum of $\phi[C]$ is the limit in the L^p norm of $\phi(\nu)$ as ν traverses the nodes in C in increasing order.*
2. *If $(C_n)_{n<\kappa}$ is a partition of S into almost norm-maximizing chains (where $\kappa \leq \omega$), then the $\preceq$-infima of $\phi[C_0], \phi[C_1], \ldots$ are disjointly supported. Furthermore, if A is an atom of Ω, then there exists a unique n such that A is the support of the $\preceq$-infimum of $\phi[C_n]$.*

The next theorem generalizes a result on isomorphisms of disintegrations from [9] and establishes an important independence property for disintegrations.

Theorem 6.20 *Suppose ϕ_j is a disintegration of $L^p(\Omega_j)$ for $j \in \{0, 1\}$ and suppose f is an isomorphism of ϕ_0 and ϕ_1. Then there is a unique isomorphism T_f of $L^p(\Omega_0)$ onto $L^p(\Omega_1)$ so that $T_f(\phi_0(\nu)) = \phi_1(f(\nu))$ for all $\nu \in \mathrm{dom}(\phi_0)$.*

6.1.6 Moduli of Convexity

We now turn to the geometry of L^p spaces, and we begin with the following classical definition.

Definition 6.21 Suppose $\mathcal{B}$ is a nonzero Banach space. For all $0 \leq \epsilon \leq 2$, let

$$\delta_{\mathcal{B}}(\epsilon) = \inf_{u,v} \left(1 - \frac{\|u + v\|}{2}\right),$$

where u, v range over all unit vectors of $\mathcal{B}$ so that $\|u - v\| \geq \epsilon$. The function $\delta_{\mathcal{B}}$ is called the *modulus of uniform convexity* of $\mathcal{B}$.

Intuitively, the modulus of uniform convexity quantifies the convexity of the unit sphere of $\mathcal{B}$. The computation of the moduli of uniform convexity for Lebesgue spaces was performed by Hanner in 1956 [2]. The following definition is required to state his result.

Definition 6.22 Suppose $1 < p < \infty$ and $0 < \epsilon \leq 2$.

1. If $1 < p \leq 2$, then let $\delta(p, \epsilon)$ denote the unique number $\delta \in [0, 1]$ so that

$$\left(1 - \delta + \frac{\epsilon}{2}\right)^p + \left|1 - \delta - \frac{\epsilon}{2}\right|^p = 2.$$

2. If $2 < p$, then set $\delta(p, \epsilon)$ to be

$$1 - \left(1 - \left(\frac{\epsilon}{2}\right)^p\right)^{1/p}.$$

We can now state Hanner's Theorem (which, as previously mentioned, he attributes to A. Beurling).

Theorem 6.23 *Suppose* $1 < p < \infty$ *and* $\mathcal{B}$ *is either* ℓ^p *or* $L^p[0, 1]$. *Then* $\delta_{\mathcal{B}}(\epsilon) = \delta(p, \epsilon)$ *whenever* $0 < \epsilon \leq 2$.

The following minor extension of Theorem 6.23 was proven in [11].

Proposition 6.24 *Suppose* $1 < p < \infty$, *and let* $\mathcal{B}$ *be a separable* L^p *space whose dimension is at least* 2. *Then the following hold:*

1. $\delta_B(\epsilon) = \delta(p, \epsilon)$ *for all* $\epsilon \in (0, 2]$.
2. *If* $0 < \epsilon < 2$,

$$\delta_{\mathcal{B}}(\epsilon) = \inf_{u,v} \left(1 - \frac{1}{2}\|u + v\|_p\right),$$

where u, v *range over all unit vectors of* $\mathcal{B}$ *so that* $\|u + v\|_p > \epsilon$.

Note that Proposition 6.24.1 follows from the classification of L^p spaces. We will also use the following, which is a fairly straightforward consequence of the definitions.

Proposition 6.25 *Fix* $0 < \epsilon < 2$.

1. *If* $2 \leq p_1 < p_2 < \infty$, *then* $\delta(p_2, \epsilon) < \delta(p_1, \epsilon)$.
2. *If* $1 < p_1 < p_2 \leq 2$, *then* $\delta(p_1, \epsilon) < \delta(p_2, \epsilon)$.

6.2 Computable Presentations

Here, we consider the following question: "Which Lebesgue spaces are computably presentable?" We begin with the following observation, which is an immediate consequence of the classification of separable L^p spaces.

Proposition 6.26 *If $p \geq 1$ is computable, then every separable L^p space is computably presentable.*

Proposition 6.26 leads naturally to the following question.

Question 6.27 Suppose $1 \leq p < \infty$. If there is a computably presentable L^p space, does it follow that p is computable?

Unfortunately, the answer to Question 6.27 is "no," and there is a simple counterexample: if $1 \leq p < \infty$, then $\ell_1^p = \mathbb{F}$ is computably presentable. We are thus led to the following modification of Question 6.27.

Question 6.28 Suppose $1 \leq p < \infty$. If there is a computably presentable L^p space whose dimension is at least 2, does it follow that p is computable?

The following is a partial result that seems to indicate that the answer to Question 6.28 is "yes."

Theorem 6.29 ([11]) *Suppose $2 \leq p < \infty$. If there is a computably presentable L^p space whose dimension is at least 2, then p is computable.*

Proof Suppose, without loss of generality, that $2 < p$. Suppose $\mathcal{B}$ is a computably presentable L^p space, and fix rational numbers r_0, r_1 so that $r_0 < p < r_1$. Suppose $k \in \mathbb{N}$ is given and we wish to compute $q \in \mathbb{Q}$ so that $|q - p| < 2^{-k}$. By means of Theorem 6.12, we can now test each rational $r \in [r_0, r_1]$ to see if $\mathcal{B}$ embeds into $L^r[0, 1]$. By Theorem 6.11, $\mathcal{B}$ embeds into $L^r[0, 1]$ if and only if $r = p$. If $r \neq p$, then we will eventually eliminate r as a candidate for p. Moreover, we will eliminate a rational open interval that contains r. After enough time has elapsed, we will have covered all of $[r_0, r_1]$ except for a subinterval whose diameter is smaller than 2^{-k}. We then choose q to be any rational number in this interval. □

When $p < 2$, the best result to date is the following:

Theorem 6.30 ([11]) *Suppose $1 \leq p < 2$. If there is a computably presentable L^p space whose dimension is at least 2, then p is right c.e.*

Proof Suppose there is a computably presentable L^p space whose dimension is at least 2. Given a rational $r < 2$, we first use Proposition 6.24.1 to calculate the modulus of convexity for a nonzero L^r space. At the same time, we use Proposition 6.24.2 to estimate the modulus from above. If we see that the calculated value is larger than the actual value, then, by Proposition 6.25, we can list r as a number that is larger than p. □

We note that due to Theorem 6.13, it is not possible to use the technique of this proof to enumerate the left Dedekind cut of p.

While Theorem 6.29 might lead one to believe that the answer to Question 6.28 is "yes," one would expect that such a positive answer would not be uniform. However, one might expect that it would at least be locally uniform; that is, given the information that the exponent is in a sufficiently small interval I, the proof should uniformly compute the exponent from a presentation. As it turns out, should the answer to Question 6.28 be positive, it would not even be locally uniform in this sense [11]:

Theorem 6.31 *Fix an open interval* $I \subseteq (1, 2)$. *Then there is no computable* $f : \mathbb{N} \to \mathbb{N}$ *with the property that if* $e \in \mathbb{N}$ *indexes a presentation of a Lebesgue space* $\mathcal{B}$ *whose exponent is in* I, *then* $f(e)$ *indexes the exponent of* $\mathcal{B}$.

In the finite-dimensional case, things work out as expected [11]:

Theorem 6.32 *If there is a computably presentable* L^p *space whose dimension is finite but larger than* 1, *then* p *is computable.*

In light of the results of this section, we make the following conjecture.

Conjecture 6.33 *If* $1 \leq p < 2$ *is right c.e., then* $L^p[0, 1]$ *is computably presentable.*

6.3 Computable Categoricity and Degrees of Categoricity

The following was first observed by Pour-El and Richards, though not in the language of computable structure theory [12].

Proposition 6.34 *Every separable* L^2 *space is computably categorical.*

Proof Suppose $\mathcal{B}^\#$ is a computable presentation of an L^2 space. By standard techniques, it is possible to show that $\mathcal{B}^\#$ has a computable Hamel basis. It follows from the polar identity that the inner product is a computable functional of $\mathcal{B}^\#$. We can then perform Gram-Schmidt orthonormalization to obtain a computable orthonormal basis. □

In contrast, Pour-El and Richards showed that ℓ^1 is not computably categorical. Their proof relies on the observation that linear isometries permute the extreme points of the unit ball. They did not address the computable categoricity of ℓ^p for computable $p \neq 2$. In 2013, Melnikov renewed interest in this question [13]. In 2015, McNicholl proved the following [14]:

Theorem 6.35 *If* $p \geq 1$ *is computable, then* ℓ^p *is computably categorical only when* $p = 2$.

The proof of the previous theorem uses Corollary 6.10 and otherwise follows the argument of Pour-El and Richards for ℓ^1. Two years later, McNicholl ascertained the degree of categoricity for ℓ^p spaces [4].

Theorem 6.36 *Suppose $p \geq 1$ is a computable real such that $p \neq 2$. Then the degree of categoricity of ℓ^p is $\mathbf{0}'$.*

Two years later, the computable categoricity of $L^p[0, 1]$ for computable $p \geq 1$ was demonstrated by Clanin et al. [9].

Theorem 6.37 *If $p \geq 1$ is a computable real, then $L^p[0, 1]$ is computably categorical.*

The degrees of categoricity for the remaining types of L^p spaces were settled a year later by Brown and McNicholl [10].

Theorem 6.38 *Suppose $p \geq 1$ is a computable real such that $p \neq 2$.*

1. *The degree of categoricity of $\ell^p \oplus_p L^p[0, 1]$ is $\mathbf{0}''$.*
2. *The degree of categoricity of $\ell_n^p \oplus_p L^p[0, 1]$ is $\mathbf{0}'$.*

The proofs of the above three theorems rely on the computability of disintegrations, which we define now. Let $\mathcal{B}^\#$ be a computable presentation of an L^p space $\mathcal{B}$, and suppose ϕ is a disintegration of $\mathcal{B}$. We say that ϕ is a *computable disintegration of $\mathcal{B}^\#$* if there is an algorithm that, given $\nu \in \mathbb{N}^*$ and k as input, does not halt if $\nu \notin S$ and otherwise produces a rational vector v of $\mathcal{B}^\#$ so that $\|v - \phi(\nu)\| < 2^{-k}$.

The inequality in Theorem 6.6 provides the means for a search procedure for disjointly supported vectors. By means of this procedure, we can demonstrate the following.

Theorem 6.39 *Suppose p is a computable real such that $p \geq 1$ and $p \neq 2$. If $\mathcal{B}^\#$ is a computable presentation of an L^p space, then there is a computable disintegration of $\mathcal{B}^\#$.*

One we have computed a disintegration, the proof of the following is fairly straightforward.

Theorem 6.40 *If $\mathcal{B}^\#$ is a computable presentation of an L^p space and ϕ is a computable disintegration of $\mathcal{B}^\#$, then there is a partition $(C_n)_{n<\kappa}$ of $\mathrm{dom}(\phi)$ into uniformly c.e. almost norm-maximizing chains.*

Theorem 6.36 is proven by using the oracle $\mathbf{0}'$ to discern which almost norm-maximizing chains have nonzero limit. These can then be treated as names of a nonzero scalar multiple of a standard basis vector, and an isomorphism onto the standard presentation of ℓ^p can be computed accordingly. Theorem 6.37 is demonstrated by first forming a disintegration of the standard presentation of $L^p[0, 1]$ that is computably isomorphic to a computable disintegration of the given nonstandard computable presentation. It is then shown that the lifting of this isomorphism given by Theorem 6.20 is computable. The proof of Theorem 6.38 proceeds by using the oracles to compute the projections onto the purely atomic and the atomless components.

6.4 Index Sets

We now apply the methodology of index sets to various classes of L^p spaces. The results in this section are proven in [15].

We begin with problems that require determining whether a given number indexes a presentation of an L^p space in a given class. In the cases of general L^p spaces and ℓ^2 and ℓ^2_n spaces, the results are unsurprising.

Theorem 6.41 *Suppose $p \geq 1$ is a computable real. Then the set of all indices of L^p space presentations is Π^0_2-complete.*

Theorem 6.42 *The set of all indices of presentations of ℓ^2 is Π^0_2-complete, as is the set of all indices of presentations of ℓ^2_n for each $n \geq 1$.*

When we consider classes of L^p spaces with $p \neq 2$, the situation becomes more interesting.

Theorem 6.43 *Let p be a computable real such that $p \geq 1$ and $p \neq 2$.*

1. *For each $n \geq 1$, the set of all indices of presentations of ℓ^p_n is Π^0_2-complete.*
2. *The set of all indices of presentations of ℓ^p is Π^0_3-complete.*
3. *The set of all indices of presentations of $L^p[0, 1]$ is Π^0_2-complete.*
4. *For each $n \geq 1$, the set of all indices of presentations of $\ell^p_n \oplus_p L^p[0, 1]$ is d-Σ^0_2-complete.*
5. *The set of all indices of presentations of $\ell^p \oplus_p L^p[0, 1]$ is d-Σ^0_3-complete.*

Now we consider the problem of determining whether a given number indexes a presentation of a Lebesgue space. The best known result along these lines is the following:

Theorem 6.44 *The set of all indices of Lebesgue space presentations is Π^0_3.*

Any attempt to prove the optimality of Theorem 6.44 encounters the same difficulty involved in the resolution of Question 6.28, namely, computing the exponent p of a Lebesgue space when $p < 2$. We nevertheless conjecture that the upper bound in Theorem 6.44 is the best possible.

Finally, we turn to the isomorphism problem for L^p spaces.

Theorem 6.45 *Let p be a computable real such that $p \geq 1$ and $p \neq 2$. Then the set of all pairs $(e, e') \in \mathbb{N}^2$ so that e and e' index presentations of isomorphic L^p spaces is co-3-Σ^0_3 complete.*

To the best of our knowledge, the complexity class in Theorem 6.45 does not occur elsewhere in computable structure theory.

References

1. Pilar Cembranos and José Mendoza. *Banach Spaces of Vector-Valued Functions*, volume 1676 of *Lecture Notes in Mathematics*. Springer-Verlag, Berlin, 1997.
2. Olof Hanner. On the uniform convexity of L^p and l^p. *Ark. Mat.*, 3:239–244, 1956.
3. John Lamperti. On the isometries of certain function-spaces. *Pacific J. Math.*, 8:459–466, 1958.
4. Timothy H. McNicholl. Computable copies of ℓ^p. *Computability*, 6(4):391–408, 2017.
5. S. Banach. *Theory of Linear Operations*, volume 38 of *North-Holland Mathematical Library*. North-Holland Publishing Co., Amsterdam, 1987. Translated from the French by F. Jellett, with comments by A. Pełczyński and Cz. Bessaga.
6. R. E. A. C. Paley. Some theorems on abstract spaces. *Bull. Amer. Math. Soc.*, 42(4):235–240, 1936.
7. Jean-Louis Krivine. Plongement des espaces normés dans les L^p pour $p>2$. *C. R. Acad. Sci. Paris*, 261:4307–4310, 1965.
8. Jean Bretagnolle, Didier Dacunha-Castelle, and Jean-Louis Krivine. Lois stables et espaces L^p. *Ann. Inst. H. Poincaré Sect. B (N.S.)*, 2:231–259, 1965/1966.
9. Joe Clanin, Timothy H. McNicholl, and Don M. Stull. Analytic computable structure theory and L^p spaces. *Fund. Math.*, 244(3):255–285, 2019.
10. Tyler Brown and Timothy H. McNicholl. Analytic computable structure theory and L^p-spaces part 2. *Arch. Math. Logic*, 59(3-4):427–443, 2020.
11. Timothy H. McNicholl. Computing the exponent of a Lebesgue space. *J. Log. Anal.*, 12:Paper No. 7, 28, 2020.
12. Marian B. Pour-El and J. Ian Richards. *Computability in Analysis and Physics*. Perspectives in Mathematical Logic. Springer-Verlag, Berlin, 1989.
13. Alexander G. Melnikov. Computably isometric spaces. *J. Symbolic Logic*, 78(4):1055–1085, 2013.
14. Timothy H. McNicholl. A note on the computable categoricity of ℓ^p spaces. In *Evolving Computability*, volume 9136 of *Lecture Notes in Comput. Sci.*, pages 268–275. Springer, Cham, 2015.
15. Tyler A. Brown, Timothy H. McNicholl, and Alexander G. Melnikov. On the complexity of classifying Lebesgue spaces. *J. Symb. Log.*, 85(3):1254–1288, 2020.

Chapter 7
Operator Algebras

In this chapter, we survey some of the known results about the effective structure theory of operator algebras, that is, C*-algebras and (tracial) von Neumann algebras. Throughout this chapter, our scalar field is $\mathbb{C}$.

7.1 Classical Background

We begin by introducing some background material on operator algebras. Szabó's article in [1] is a more thorough (but still fairly short) introduction to C*-algebras, and Ioana's introduction to von Neumann algebras in [1] is a more expansive discussion of von Neumann algebras.

7.1.1 What Is a C-Algebra? Abstract and Concrete Definitions*

There are two main ways to introduce the topic of C*-algebras. One is commonly regarded as abstract, while the other is often viewed as concrete. We first describe the abstract approach wherein a C*-algebra is defined as a kind of structure that satisfies certain conditions. We begin with the definition of a $*$-algebra.

Definition 7.1 A *$*$-algebra* consists of an algebra $\mathcal{A}$ over $\mathbb{C}$[1] together with an involution $x \mapsto x^*$ that satisfies the following conditions:

---

[1] There is such a thing as a *real* C*-*algebra*, but we will rarely have occasion to mention them and thus restrict our discussion to complex scalars.

J. N. Y. Franklin et al., *Effective Metric Structure Theory*, Springer Monographs in Mathematics, https://doi.org/10.1007/978-3-032-13322-9_7

1. For all $x, y \in \mathcal{A}$, $(x + y)^* = x^* + y^*$.
2. For all $x, y \in \mathcal{A}$, $(xy)^* = y^* x^*$.
3. For all $x \in \mathcal{A}$ and $\lambda \in \mathbb{C}$, $(\lambda x)^* = \bar{\lambda} x^*$.
4. If $\mathcal{A}$ is unital, then $\mathbf{1}_{\mathcal{A}}^* = \mathbf{1}_{\mathcal{A}}$.

If $\mathcal{A}$ is a $*$-algebra and if $a \in \mathcal{A}$, then we call a^* the *adjoint* of a. There are obvious notions of a $*$-subalgebra of a $*$-algebra and a unital $*$-subalgebra of a unital $*$-algebra.

A $*$-*homomorphism* between $*$-algebras is an algebra homomorphism that also preserves the adjoint operation. If $\varphi : \mathcal{A} \to \mathcal{B}$ is a $*$-homomorphism between unital $*$-algebras, then we also require that φ maps the unit of $\mathcal{A}$ to the unit of $\mathcal{B}$.

A basic yet interesting example of a $*$-algebra is $M_n(\mathbb{C})$ with the usual adjoint operation. This is an instance of a more general example which is for us the most relevant, namely $\mathcal{B}(\mathcal{H})$, the $*$-algebra of bounded operators on the Hilbert space $\mathcal{H}$ and its $*$-subalgebras. Recall that for a Hilbert space $\mathcal{H}$, a linear operator $T : \mathcal{H} \to \mathcal{H}$ is *bounded* if its *operator norm* $\|T\| = \sup\{\|T\xi\| : \|\xi\| \leq 1\}$ is finite. It is easily seen that $\mathcal{B}(\mathcal{H})$ is a unital algebra whose unit is the identity map $\mathrm{Id}_{\mathcal{H}}$ and whose operations are addition, composition, and scalar multiplication. Recall also that when $T \in B(\mathcal{H})$, its *adjoint* is the unique $T^* \in \mathcal{B}(\mathcal{H})$ for which $\langle T\xi, \eta\rangle = \langle \xi, T^*\eta\rangle$ for all $\xi, \eta \in \mathcal{H}$. It is routine to verify that $T \mapsto T^*$ satisfies all the conditions in Definition 7.1. Thus, $B(\mathcal{H})$ is a unital $*$-algebra.

A $*$-algebra becomes a C*-algebra when it is equipped with a complete C*-norm, which we define now.

Definition 7.2 Suppose that $\mathcal{A}$ is a $*$-algebra. A C*-*norm* on $\mathcal{A}$ is a norm on $\mathcal{A}$ that satisfies the following conditions:

1. For all $x, y \in \mathcal{A}$, $\|xy\| \leq \|x\|\|y\|$.
2. For every $x \in \mathcal{A}$, $\|x^*\| = \|x\|$.
3. If $\mathcal{A}$ is unital, then $\|\mathbf{1}_{\mathcal{A}}\| = 1$.
4. For every $x \in \mathcal{A}$, $\|x^* x\| = \|x\|^2$.

The first conditions of Definition 7.2 are the usual axioms for defining a (unital) *normed* $*$-algebra. The last condition is known as the C*-*identity*, and it is what makes a C*-norm a C*-norm.

We now state the abstract definition of the class of C*-algebras.

Definition 7.3 A C*-*algebra* is a $*$-algebra that is equipped with a complete C*-norm.

One natural example of a C*-algebra is $\mathrm{C}(X; \mathbb{C})$, augmented with pointwise multiplication and the adjoint defined by conjugation (that is, $f^*(p) = \overline{f(p)}$). *Throughout this chapter, we denote this algebra simply by $C(X)$ and we caution the reader that this differs from conventions used in earlier chapters.* Another natural example is $M_n(\mathbb{C})$ with the operator norm.

The completeness requirement of Definition 7.3 is important, but not especially onerous. In particular, if $\mathcal{A}$ is a $*$-algebra equipped with a C*-norm, then the $*$-algebra operations and norm extend naturally to the completion of the $*$-algebra, which is then a C*-algebra.

An important consequence of the C*-identity is that every $*$-homomorphism between C*-algebras is 1-Lipschitz. It then follows that a $*$-homomorphism between C*-algebras is an isometric embedding if and only if it is injective. This entails that *for every $*$-algebra $\mathcal{A}$, there is at most one* C**-norm on $\mathcal{A}$.*

We now turn to the concrete view of C*-algebras. It is an easy exercise to see that the operator norm on $\mathcal{B}(\mathcal{H})$ is a C*-norm. Thus every complete $*$-subalgebra of $B(\mathcal{H})$ is a C*-algebra; such C*-algebras are called *concrete* C**-algebras*. The *Gelfand-Naimark Theorem* states that every C*- algebra is $*$-isomorphic to a concrete C*-algebra.

Both perspectives on C*-algebras are useful. In particular, we will apply terminology for bounded linear operators to elements of arbitrary C*-algebras. For example, a point p of a C*-algebra is a *projection* if it is idempotent and self adjoint.

7.1.2 What Is a von Neumann Algebra?

When introducing the class of von Neumann algebras, we once again have a choice between concrete and abstract approaches. The concrete is perhaps the easier of the two to grasp. Accordingly, we define a *concrete von Neumann algebra* to be a unital $*$-subalgebra of $\mathcal{B}(\mathcal{H})$ that is closed in the *weak operator topology* (WOT). This is the smallest topology on $\mathcal{B}(\mathcal{H})$ for which the map $T \mapsto \langle T\xi, \eta\rangle$ is continuous for all $\xi, \eta \in \mathcal{H}$. A subbasic open neighborhood of $T \in \mathcal{B}(\mathcal{H})$ in the WOT has the form $\{S \in \mathcal{B}(\mathcal{H}) : |\langle T\xi, \eta\rangle - \langle S\xi, \eta\rangle| < \epsilon\}$ for some $\xi, \eta \in \mathcal{H}$ and some $\epsilon > 0$. It is easy to check that the WOT is a weaker topology on $\mathcal{B}(\mathcal{H})$ than the operator norm topology. In other words, every sequence of operators that converges in the operator norm also converges in the WOT. Thus, every concrete von Neumann algebra is a (unital) concrete C*-algebra. In general, von Neumann algebras are much "bigger" than general C*-algebras due to the, well, weakness of the WOT.

A theorem of Sakai provides an abstract reformulation: a unital C*-algebra $\mathcal{M}$ is $*$-isomorphic to a concrete von Neumann algebra if and only if $\mathcal{M}$ is, as a Banach space, linearly isometric to the dual of another Banach space $\mathcal{B}$. Accordingly, a Banach space $\mathcal{B}$ is called the *predual* of $\mathcal{M}$ if $\mathcal{M}$ is linearly isometric to $\mathcal{B}^*$. Sakai's Theorem also states that, up to linear isometry, $\mathcal{M}$ has exactly one predual. Some authors refer to such *abstract von Neumann algebras* as W**-algebras*, although we will refrain from doing so.

Von Neumann's *Bicommutant Theorem* provides a purely algebraic definition of von Neumann algebras that is incredibly important to the theory. First, given a subset $\mathcal{S} \subseteq \mathcal{B}(\mathcal{H})$, define $\mathcal{S}'$ to be $\{T \in \mathcal{B}(\mathcal{H}) : TS = ST \text{ for all } S \in \mathcal{S}\}$. $\mathcal{S}'$ is known as the *commutant of* $\mathcal{S}$. Note that for any $\mathcal{S} \subseteq \mathcal{B}(\mathcal{H})$ for which $\mathcal{S}^* = \mathcal{S}$, $\mathcal{S}'$ is a von Neumann subalgebra of $\mathcal{B}(\mathcal{H})$ and $\mathcal{S} \subseteq \mathcal{S}'' := (\mathcal{S}')'$. Von Neumann's Bicommutant Theorem states that for any unital $*$-subalgebra $\mathcal{A}$ of $\mathcal{B}(H)$, $\mathcal{A}''$ coincides with the WOT-closure of $\mathcal{A}$ in $\mathcal{B}(\mathcal{H})$. This common algebra is the von Neumann algebra generated by $\mathcal{A}$. In fact, the Bicommutant Theorem shows that both of these coincide with the closure of $\mathcal{A}$ in the *strong operator topology* (SOT), which is the topology of pointwise convergence. A subbasic open neighborhood of $T \in \mathcal{B}(\mathcal{H})$ in the SOT has the form $\{S \in \mathcal{B}(\mathcal{H}) : \|(T-S)\xi\| < \epsilon\}$ for some $\xi \in \mathcal{H}$ and some $\epsilon > 0$. A consequence of the Bicommutant Theorem is that a concrete C*-algebra $\mathcal{M}$ is a von Neumann algebra if and only if $\mathcal{M} = \mathcal{M}''$.

7.1.3 Additional Vocabulary and Results for Operator Algebras

While nonunital C*-algebras exist (consider ℓ^1 with the multiplication given by convolution and the adjoint given by conjugation), by a well-known construction, every such algebra is $*$-isomorphic to a subalgebra of a unital C*-algebra. Therefore, *unless stated otherwise, in the rest of this chapter, we restrict attention to unital* C*-*algebras*. This convention will allow us to avoid any thorny issues that often arise in the nonunital context.

A C*-algebra is *commutative* (or *abelian*) if its multiplication is commutative. An example of such an algebra is $C(X)$. In fact, all (unital) commutative C*-algebras are of this form, and there is a dual equivalence of categories (known as *Gelfand duality*) between compact Hausdorff spaces with continuous maps and (unital) commutative C*-algebras with $*$-homomorphisms. For this reason, C*-algebra theory is often dubbed "noncommutative topology." There is a similar duality for nonunital C*-algebras and locally compact Hausdorff spaces.

An *ideal* of a C*-algebra $\mathcal{A}$ is a closed subspace of $\mathcal{A}$ that is closed under both left and right multiplication by elements of $\mathcal{A}$. It can be shown that ideals are also closed under the adjoint. $\mathcal{A}$ is called *simple* if it has no ideals other than $\{0\}$ and $\mathcal{A}$.

A linear functional φ on a C*-algebra $\mathcal{A}$ is a *state* if $\varphi(\mathbf{1}_{\mathcal{A}}) = 1$ and if $\varphi(a^*a) \geq 0$ for all $a \in \mathcal{A}$ (that is, φ is *positive*). If φ is a state on $\mathcal{A}$ so that $\varphi(a^*a) > 0$ when $a \neq 0$, then we say that φ is *faithful*. As an example of a state, fix a vector ξ in a Hilbert space $\mathcal{H}$ and consider the map $\varphi_\xi : \mathcal{B}(\mathcal{H}) \to \mathbb{C}$ given by $\varphi_\xi(T) = \langle T\xi, \xi \rangle$. It is easily seen that φ_ξ is a positive linear functional. If, in addition, $\|\xi\| = 1$, then $\varphi_\xi(\mathrm{Id}_{\mathcal{H}}) = 1$ and so φ_ξ is a state. In case $\mathcal{A} \subseteq \mathcal{B}(\mathcal{H})$ is a concrete unital C*-algebra and $\varphi = \varphi_\xi|_{\mathcal{A}}$ for some unit vector $\xi \in \mathcal{H}$, we say that φ is a *vector state* on $\mathcal{A}$.

Before proceeding with additional background on von Neumann algebras, we outline a proof of the Gelfand-Naimark Theorem. This result can be reformulated in terms of representations of C*-algebras. Given a C*-algebra $\mathcal{A}$, a *representation* of $\mathcal{A}$ is a $*$-homomorphism $\pi : \mathcal{A} \to \mathcal{B}(\mathcal{H})$ for some Hilbert space $\mathcal{H}$. To avoid trivialities, it is required that $\{\pi(a)(\xi) : a \in \mathcal{A}, \xi \in \mathcal{H}\}$ is dense in $\mathcal{H}$. If $\mathcal{A}$ is unital, this condition is equivalent to assuming that $\pi(\mathbf{1}_{\mathcal{A}}) = \mathrm{Id}_{\mathcal{H}}$. The representation π is *faithful* if it is moreover injective (equivalently isometric). Thus, the Gelfand-Naimark theorem states that every C*-algebra admits a faithful representation.

The key to building a faithful representation of a C*-algebra is the *Gelfand-Naimark-Segal* (or *GNS*) construction, which proceeds as follows. Fix a C*-algebra $\mathcal{A}$. Suppose that φ is a state on $\mathcal{A}$. One can then define the sesquilinear form $\langle \cdot, \cdot \rangle_\varphi$ on $\mathcal{A}$ by setting $\langle x, y\rangle_\varphi = \varphi(y^* x)$. This form induces the seminorm $\|x\|_\varphi = \sqrt{\langle x, x\rangle_\varphi} = \sqrt{\varphi(x^* x)}$. If φ is not faithful, then $\langle \cdot, \cdot \rangle_\varphi$ need not be an inner product due to the existence of nonzero x for which $\langle x, x\rangle_\varphi = 0$. We remedy this by defining $\mathcal{H}_\varphi$ to be the completion of the inner product space obtained from $\mathcal{A}$ after quotienting out by the elements x with $\|x\|_\varphi = 0$. For $x \in \mathcal{A}$, we denote the equivalence class of x in $\mathcal{H}_\varphi$ by $\hat{x}$. We then consider the map $\pi_\varphi : \mathcal{A} \to \mathcal{B}(\mathcal{H}_\varphi)$ that is uniquely defined by the condition $\pi_\varphi(x)(\hat{y}) = \widehat{xy}$. It is straightforward to check that π_φ is a representation of $\mathcal{A}$, called the *GNS respresentation* of $\mathcal{A}$ with respect to φ. Moreover, π_φ is faithful precisely when φ is faithful. The key to the proof of the Gelfand-Naimark construction is to show that $\mathcal{A}$ has enough states so that the direct sum of the various GNS representations is faithful.

We now set forth some information on von Neumann algebras. A von Neumann algebra $\mathcal{M}$ is called *separable* if its predual is a separable Banach space. This is equivalent to having a concrete representation of $\mathcal{M}$ on a separable Hilbert space. Hence, one sometimes calls such a von Neumann algebra *separably acting*.

Just as in the case of C*-algebras, we can completely characterize the commutative von Neumann algebras as follows. Fix a measure space Ω. For each $f \in L^\infty(\Omega)$ and $g \in L^2(\Omega)$, let $M_f(g) = fg$. We then view $L^\infty(\Omega)$ as a subspace of $\mathcal{B}(L^2(\Omega))$ by identifying $f \in L^\infty(\Omega)$ with M_f. It is routine to check that $L^\infty(\Omega) = L^\infty(\Omega)'$, whence $L^\infty(\Omega)$ is a commutative von Neumann algebra. Moreover, all commutative von Neumann algebras have this form. For this reason, von Neumann algebra theory is often dubbed "noncommutative measure theory."

Just as with C*-algebras, $*$-homomorphisms between von Neumann algebras are automatically 1-Lipschitz. However, we use a continuity condition that relates to the WOT since this is the topology used to define von Neumann algebras. More precisely, a positive linear map $\varphi : \mathcal{M} \to \mathcal{N}$ between von Neumann algebras is *normal* if the restriction of φ to the operator norm unit ball of $\mathcal{M}$ is continuous when $\mathcal{M}$ and $\mathcal{N}$ are equipped with their weak operator topologies. The restriction to the operator norm unit ball may seem slightly unsightly. However, this is equivalent to saying that φ is continuous under the weak*-topologies on $\mathcal{M}$ and $\mathcal{N}$ when they are viewed as dual Banach spaces. One can reformulate normality in an intrinsic way that does not refer to the particular realization of $\mathcal{M}$ and $\mathcal{N}$, namely $\varphi : \mathcal{M} \to \mathcal{N}$ is normal if and only if it is positive, linear, and $\varphi(\sup_{i\in I} x_i) = \sup_{i\in I} \varphi(x_i)$ for every bounded increasing net $(x_i)_{i\in I}$ of positive elements in $\mathcal{M}$.

Most of the effective structure theory for von Neumann algebras focuses on the class of *tracial von Neumann algebras*, which we now introduce. A state φ on $\mathcal{A}$ is called a *tracial state* if $\varphi(ab) = \varphi(ba)$ for all $a, b \in \mathcal{A}$. For example, the normalized trace tr on $M_n(\mathbb{C})$ given by $\mathrm{tr}(a) = \frac{1}{n}\mathrm{Tr}(a)$ is a tracial state on $M_n(\mathbb{C})$ (where Tr is the usual "sum of the diagonal" trace on $M_n(\mathbb{C})$). Since $\mathbb{C}$ is a von Neumann algebra, it makes sense to speak of *normal* states on von Neumann algebras. A faithful normal tracial state on a von Neumann algebra is simply referred to as a *trace*. A von

Neumann algebra is called *finite* if it admits a trace. (This terminology makes much more sense if you introduce Murray-von Neumann equivalence of projections.) A *tracial von Neumann algebra* is a pair $(\mathcal{M}, \tau)$ where $\mathcal{M}$ is a von Neumann algebra and τ is a trace on $\mathcal{M}$. An embedding of tracial von Neumann algebras is a normal, injective $*$-homomorphism that preserves the trace. The normalized trace tr on $M_n(\mathbb{C})$ is a trace (in the von Neumann algebra sense) on $M_n(\mathbb{C})$. However, when $\mathcal{H}$ is infinite dimensional, there is no trace on $\mathcal{B}(\mathcal{H})$.

The trace on a von Neumann algebra may not be unique. However, there is an important algebraic condition that guarantees the uniqueness of the trace. A von Neumann algebra M is called a *factor* if its *center* $Z(M) = \{x \in M : xy = yx\} = M \cap M'$ is trivial; that is, if $Z(M) = \mathbb{C} \cdot 1$. It is a fact that a factor admits at most one trace. Conversely, a von Neumann algebra that admits exactly one trace is a factor. The finite-dimensional tracial factors are precisely those of the form $M_n(\mathbb{C})$. An infinite-dimensional tracial factor is called a *II_1 factor*.

If (M, τ) is a tracial von Neumann algebra, then one can perform the GNS construction with respect to the trace. In this case, the associated Hilbert space is often denoted $L^2(M, \tau)$, or simply $L^2(M)$ when the trace is clear from context (such as when M is a tracial factor). The associated representation $\pi_\tau : M \to \mathcal{B}(L^2(M, \tau))$ can be shown to be a faithful, normal representation of M. This representation is called the *standard representation of M*. One can also check that, on operator norm bounded balls of M, the topology induced by $\|\cdot\|_\tau$ (when M is viewed as a subset of $L^2(M)$ via $x \mapsto \hat{x}$) is precisely the SOT.

7.1.4 How to Build an Operator Algebra

We now describe some standard methods for constructing C*-algebras.

Direct Sums

Given a family $(\mathcal{A}_i)_{i\in I}$ of C*-algebras, their Banach space direct sum $\bigoplus_{i\in I} \mathcal{A}_i$ is naturally a C*-algebra in which the adjoint is defined coordinatewise. Note that the norm is the supremum norm; that is, $\|(a_i)_{i\in I}\| = \sup_{i\in I} \|a_i\|$.

It now follows that for any finite sequence $n_1, \ldots, n_r$ of positive integers, $M_{n_1}(\mathbb{C}) \oplus \cdots \oplus M_{n_r}(\mathbb{C})$ is a finite-dimensional C*-algebra. It turns out that every finite-dimensional C*-algebra is of this form and this representation is unique up to permutation of the direct summands.

Inductive Limits

It is well known that the category of C*-algebras is closed under inductive limits; see, for example, [2, Sect. 6.2]. A C*-algebra A is called *approximately finite dimensional*

(AF) if it is an inductive limit of finite-dimensional C*-algebras. AF algebras can be completely classified by discrete, algebraic data known as the K-theory of the algebra (more specifically, the partially ordered group $K_0(A)$ together with a distinguished positive element).

A special class of AF algebras is the *uniformly hyperfinite* (*UHF*) algebras, which are inductive limits of matrix algebras. For a UHF algebra $\mathcal{A}$, its isomorphism type is completely determined by its so-called *supernatural number*, which is the formal infinite product $n_{\mathcal{A}} = \prod_p p^{n(\mathcal{A},p)}$ where the product is over all primes and $n(\mathcal{A}, p)$ is the supremum of all positive integers k such that $M_{p^k}(\mathbb{C})$ embeds into $\mathcal{A}$. Note that $n(\mathcal{A}, p)$ can be infinite. The unique UHF algebra with $n(A, p) = \infty$ for all primes p is known as the *universal UHF algebra* and is denoted by $\mathcal{Q}$. Another important case is when $n(\mathcal{A}, 2) = \infty$ and $n(\mathcal{A}, p) = 0$ for $p > 2$; this algebra is called the *CAR algebra* and is denoted by $M_{2^\infty}(\mathbb{C})$.

Universal C*-Algebras

In group theory, it is common to define a group by a set of generators and a set of relations. For example, the Klein 4-group and the dihedral group can be defined this way. In 1985, Blackadar developed a similar method for defining C*-algebras, which we now describe [3].

Let $\mathcal{G}$ be a set of noncommuting indeterminates, which we call *generators*. A *relation* for $\mathcal{G}$ is an inequality of the form $\|p(x_1, \ldots, x_n)\| \leq a$ where p is a $*$-polynomial in n noncommuting variables with no constant term, $x_1, \ldots, x_n \in \mathcal{G}$, and a is a nonnegative real number. We sometimes abuse notation and write $p(\bar{x}) = 0$ for the relation $\|p(\bar{x})\| \leq 0$. Note also that any relation $\|p(\bar{x})\| \leq a$ can always be rewritten in terms of vanishing of polynomials by declaring $p(\bar{x}) - y = 0$ and $\|y\| \leq a$.

Suppose $\mathcal{R}$ is a set of relations for $\mathcal{G}$ such that, for every generator $x \in \mathcal{G}$, there is a relation of the form $\|x\| \leq M$ in $\mathcal{R}$. We call $(\mathcal{G}, \mathcal{R})$ a *representing pair*. A *representation* of $(\mathcal{G}, \mathcal{R})$ is a pair $(\mathcal{A}, j)$ in which $\mathcal{A}$ is a C*-algebra and $j : \mathcal{G} \to \mathcal{A}$ satisfies $\|p(j(x_1), \ldots, j(x_n))\| \leq a$ for every relation $\|p(x_1, \ldots, x_n)\| \leq a$ in $\mathcal{R}$. A representation $(\mathcal{A}, \iota)$ of $(\mathcal{G}, \mathcal{R})$ is *universal* if for every representation $(\mathcal{B}, j)$ of $(\mathcal{G}, \mathcal{R})$ there is a unique $*$-homomorphism $\varphi : \mathcal{A} \to \mathcal{B}$ so that $\varphi(\iota(x)) = j(x)$ for all $x \in \mathcal{G}$. We note that a universal representation $(\mathcal{A}, \iota)$ of $(\mathcal{G}, \mathcal{R})$ always exists,[2] that $\mathcal{A}$ is generated by $\mathrm{ran}(\iota)$, and that the universal algebra is unique up to $*$-isomorphism. We call $\mathcal{A}$ the *universal* C*-*algebra* of $(\mathcal{G}, \mathcal{R})$, and we denote it by $\mathrm{C}^*\langle\mathcal{G}|\mathcal{R}\rangle$.

When our generators have been arranged into a sequence $\bar{x}$, we may write $\mathrm{C}^*\langle\bar{x}|\mathcal{R}\rangle$ instead of $\mathrm{C}^*\langle\bar{G}|\mathcal{R}\rangle$. Given that we remain in the context of unital C*-algebras

[2] This is a consequence of the fact that we included a norm bound for each generator. There are conditions that guarantee the existence of the universal algebra for a given representing pair without including a norm bound on each generator; see Blackadar [3].

throughout this chapter, we implicitly assume that we have a distinguished generator for the unit and include relations stating that it is a self-adjoint idempotent that acts as a multiplicative identity.

A C*-algebra $\mathcal{A}$ is *finitely presented* if it is the universal C*-algebra of a finite set of relations (in which case the set of generators must be finite as well).

We now use universal C*-algebras to define a number of important examples. We begin with the contractive C*-algebras.

Definition 7.4 For each positive integer n, set

$$\mathcal{F}(n) = C^*\langle x_1, \ldots, x_n \mid \|x_1\| \leq 1, \ldots, \|x_n\| \leq 1\rangle.$$

Similarly, we define $\mathcal{F}(\omega)$ to be $C^*\langle (x_j)_{j\in\mathbb{N}} \mid \{\|x_j\| \leq 1 : j \in \mathbb{N}\}\rangle$. When $0 < n \leq \omega$, $\mathcal{F}(n)$ is called a *universal contraction algebra*.

Next, we use universal algebras to define the Toeplitz algebras.

Definition 7.5 The *Toeplitz algebra* $\mathcal{T}$ is $C^*\langle x \mid \|x\| \leq 1, \|x^*x - \mathbf{1}\| = 0\rangle$.

Put another way, the Toeplitz algebra is the universal C*-algebra generated by a single isometry. This is a somewhat abstract definition, but a natural concrete realization is possible. In order to motivate this realization, let us recall that when $\mathcal{H}$ is a Hilbert space, a map $T \in B(\mathcal{H})$ is a *proper isometry* if $T^*T = \mathrm{Id}_{\mathcal{H}}$ but $TT^* \neq \mathrm{Id}_{\mathcal{H}}$. Note that, by the universal property and the fact that proper isometries exist, the generator of the Toeplitz algebra must be proper. An important example of a proper isometry is the right shift $S : \ell^2 \to \ell^2$ given by $S(e_n) = e_{n+1}$. By a theorem of Colburn [4], the Toeplitz algebra is $*$-isomorphic to the $*$-subalgebra of $\mathcal{B}(\ell^2)$ generated by S.

Suppose $\mathcal{H}$ is a separable Hilbert space with orthonormal basis $(e_n)_{n\in\mathbb{N}}$. Consider the proper isometries $a_1, a_2 \in \mathcal{B}(\mathcal{H})$ determined by $a_1(e_n) = e_{2n}$ and $a_1(e_n) = e_{2n+1}$. Then the projections $a_1a_1^*$ and $a_2a_2^*$ are orthogonal and sum to the identity. We now define a universal algebra that satisfies a more general version of these relations.

Definition 7.6 Fix an integer $n \geq 2$. The *Cuntz algebra* $\mathcal{O}_n$ is the universal C*-algebra defined by n generators $a_1, \ldots, a_n$ and the following relations:

1. $\|a_i\| \leq 1$.
2. $a_j^*a_i = \delta_{ij}\mathbf{1}$.
3. $\sum_{i=1}^n a_ia_i^* = \mathbf{1}$.

We also define the Cuntz algebra $\mathcal{O}_\infty$ to be the universal C*-algebra generated by a sequence $(a_n)_{n\in\mathbb{N}}$ of contractions satisfying $a_j^*a_i = \delta_{ij}\mathbf{1}$.

The Cuntz algebras play a central role in modern C*-algebra theory. For example, a landmark result of Kirchberg shows that every separable *nuclear* C*-algebra embeds into $\mathcal{O}_2$.

Finally, we define a class of algebras that plays a prominent role in the field of noncommutative geometry.

Definition 7.7 Given any irrational number $\theta \in (0, 1)$, the *irrational rotation algebra* A_θ is the universal C*-algebra generated by two unitaries u and v that satisfy $uv = e^{2i\pi\theta}vu$.

Algebras from Groups

An important source of C*-algebras comes from groups. In fact, there are two (in general distinct) constructions of a C*-algebra from a group. To describe one of them, we need to recall the notion of the left regular representation of a group.

Suppose that G is a group. Let $\ell^2(G)$ be the Hilbert space formally generated by an orthonormal basis δ_h for all $h \in G$. Alternatively, we may view $\ell^2(G)$ as $L^2(G, \mu)$ where μ is the counting measure on G. For any $g \in G$, define u_g to be the linear operator on $\ell^2(G)$ that satisfies $u_g(\delta_h) = \delta_{gh}$ for all $h \in G$. Put another way, for each $f \in L^2(G, \mu)$ and $g, h \in G$, $u_g(f)(h) = f(g^{-1}h)$. Notice that u_g is unitary for all $g \in G$ (since $u_g^* = u_g^{-1} = u_{g^{-1}}$). Thus, the map $\lambda : G \to \mathcal{B}(\ell^2(G))$ given by $\lambda(g) = u_g$ is a unitary representation of G. This representation is called the *left regular representation* of G.

Define $\mathbb{C}[G]$ to be the set of all formal linear combinations $\sum_{g\in G} c_g g$ with only finitely many nonzero coefficients. There is a natural $*$-algebra structure on $\mathbb{C}[G]$, namely, the addition and multiplication being the obvious ones and the $*$-operation being given by $(\sum_{g\in G} c_g g)^* = \sum_{g\in G} \overline{c_g} g^{-1}$. Accordingly, we call $\mathbb{C}[G]$ the *group algebra* of G. We note that $\mathbb{C}[G]$ is unital and that the unit is the identity of the group, which we denote by e. The left regular representation λ of G extends by linearity to a unital $*$-algebra homomorphism $\pi : \mathbb{C}[G] \to B(\ell^2(G))$. By means of this extension, the group algebra can be extended to a C*-algebra, which we define now.

Definition 7.8 The *reduced group* C*-*algebra of* G, denoted $C_r^*(G)$, is the closure of $\pi(\mathbb{C}[G])$ in the operator norm topology on $\mathcal{B}(\ell^2(G))$.

For later developments, it will be important to know that there is a trace on $C_r^*(G)$ that is given by $\tau(x) = \langle x\delta_e, \delta_e\rangle$. For $x = \sum_{g\in G} c_g g \in \mathbb{C}[G]$, note that $\tau(x) = c_e$.

As mentioned earlier, there is a second C*-algebra one can construct from a group. We define this algebra now.

Definition 7.9 Given a group G, let $C^*(G)$ be the universal C*-algebra defined by using the elements of G as generators and the following set of relations:

1. $\|x\| \leq 1$ for all $x \in G$.
2. $x^*x = xx^* = e$ for all $x \in G$.
3. $xy = z$ for all $x, y, z \in G$ for which $xy = z$.

$C^*(G)$ is called the *universal group* C*-*algebra of* G.

By the universal property of universal C*-algebras, every unitary group representation $\pi : G \to B(\mathcal{H})$ extends to a $*$-homomorphism $\alpha : C^*(G) \to B(\mathcal{H})$. In particular, taking $\pi = \lambda$, the left regular representation, there is a canonical surjective $*$-homomorphism from $C^*(G)$ to $C^*_r(G)$. This canonical $*$-homomorphism is an isomorphism if and only if G is amenable.

Groups also give rise to an interesting example of tracial von Neumann algebras. Fix a group G and let $\lambda : G \to \mathcal{B}(\ell^2(G))$ denote its left regular representation. Let $\pi : \mathbb{C}[G] \to \mathcal{B}(\ell^2(G))$ be the extension of λ defined above. We then define the *group von Neumann algebra* $L(G)$ to be the von Neumann subalgebra of $\mathcal{B}(\ell^2(G))$ generated by $\pi(\mathbb{C}[G])$. Thus, $L(G) = \pi(\mathbb{C}[G])'' = \overline{\pi(\mathbb{C}[G])}^{\mathrm{SOT}} = \overline{\pi(\mathbb{C}[G])}^{\mathrm{WOT}}$. There is then a natural trace τ on $L(G)$ given by the vector state δ_e; that is to say, $\tau(x) = \langle x\delta_e, \delta_e\rangle$ for all $x \in L(G)$.

In the case of group von Neumann algebras, $L(G)$ is a factor precisely when G is an *ICC group*; that is, when all nontrivial conjugacy classes of G are infinite. If G is also infinite, then $L(G)$ is a II_1 factor. A fundamental result of Connes [5] states that all infinite ICC amenable groups have the same associated group von Neumann algebra, called the *hyperfinite II_1 factor*, which is denoted $\mathcal{R}$.

This completes our summary of the classical theory of operator algebras. We now turn to their computable presentability.

7.2 Computable Presentability of Operator Algebras

7.2.1 *Preliminary Remarks*

Suppose $\mathcal{A}$ is a C^*-algebra. We view $\mathcal{A}$ as a metric structure in the language of Banach $*$-algebras.[3] (See Example 2.8.) Hence, we are now set up to study presentations of C^*-algebras. We begin with some preliminary simple observations.

Fix a presentation $\mathcal{A}^\#$ of $\mathcal{A}$. It follows that the rational points of $\mathcal{A}^\#$ are precisely the points of the form $p(s_1, \ldots, s_n)$ where $s_1, \ldots, s_n$ are distinguished points of $\mathcal{A}^\#$ and p is a rational $*$-polynomial with no constant term. The computability of $\mathcal{A}^\#$ is equivalent to the computability of the norm on the rational points of $\mathcal{A}^\#$. Similarly, $\mathcal{A}^\#$ is right c.e. if the norm is an upper semicomputable function on the rational points, and $\mathcal{A}^\#$ is left c.e. if the norm is lower semicomputable on the rational points.

It is worth commenting that, although we have restricted attention to unital C^*-algebras, since we are treating C^*-algebras as structures in the language of Banach $*$-algebras, there is no requirement that the identity of the C^*-algebra be a rational point of the presentation. Indeed, there is not even a requirement that it is a computable point of a presentation. One could alternatively treat unital C^*-algebras as structures in the language of unital Banach $*$-algebras, in which case the identity would necessarily be a rational point of the presentation. For most of what we discuss below, there is no real difference in the choice of language. However, sometimes, especially regarding issues of uniformity, the choice of language is important [7, 8].

[3] It is a fact that the class of (dissections of) C^*-algebras forms an elementary class in this language [6].

If $\mathcal{A}_0^\#$ and $\mathcal{A}_1^\#$ are presentations of C*-algebras, then there is a natural presentation $\mathcal{A}_0^\# \oplus \mathcal{A}_1^\#$ of $\mathcal{A}_0 \oplus \mathcal{A}_1$. This presentation is computable if $\mathcal{A}_0^\#$ and $\mathcal{A}_1^\#$ are computable. The category of computable presentations of C*-algebras is also closed under computable inductive limits. A precise statement of this principle and its proof can be found in [9].

Finally, we wish to connect some classical terminology with the terminology inherited from Chap. 3. In the classical literature, when a representing pair $(\mathcal{G}, \mathcal{R})$ defines a universal C*-algebra $\mathcal{A}$, $(\mathcal{G}, \mathcal{R})$ is sometimes called a presentation of $\mathcal{A}$. We have eschewed this terminology for obvious reasons, but we now note there is a natural connection between representing pairs and presentations. On the one hand, suppose $\mathcal{A} = C^*\langle\mathcal{G}|\mathcal{R}\rangle$ and assume $\mathcal{G}$ is countable. Let ν be any surjection of $\mathbb{N}$ onto $\mathcal{G}$, and let ι be any map from $\mathcal{G}$ into $\mathcal{A}$ so that $(\mathcal{A}, \iota)$ is a universal representation of $(\mathcal{G}, \mathcal{R})$. We can then define a presentation $\mathcal{A}^\#$ of $\mathcal{A}$ by declaring its nth distinguished point to be $\iota(\nu(n))$.

Conversely, suppose $\mathcal{A}^\#$ is a presentation of $\mathcal{A}$ and let a_n denote the nth distinguished point of $\mathcal{A}^\#$. Set $\mathcal{G} = \mathbb{N}$ and declare $\|p(n_1, \ldots, n_k)\| \leq a$ to be an element of $\mathcal{R}$ just in case $\left\|p(a_{n_1}, \ldots, a_{n_k})\right\|_{\mathcal{A}} \leq a$. It then follows that $\mathcal{A} = C^*\langle\mathcal{G}|\mathcal{R}\rangle$.

In summary, presentations and universal C*-algebras are naturally related and are almost the same concept. As we shall see below, this simple insight turns out to be very valuable for computability-theoretic purposes.

7.2.2 *Matrix Algebras*

We begin our exploration of presentations of C*-algebras with the simple case of $M_n(\mathbb{C})$. Let $(E^n_{r,s})_{r,s\in\mathbb{N}}$ denote the standard system of matrix units for $M_n(\mathbb{C})$. The standard presentation of $M_n(\mathbb{C})$ is "the" presentation whose distinguished points are precisely these matrix units. To be more precise, we fix a standard computable surjection ν of $\mathbb{N}$ onto $\{1, \ldots, n\}^2$ and then declare the kth distinguished point to be $E^n_{\nu(k)}$.

The computability of this presentation is a fairly straightforward exercise in the principles of computable analysis. Fix a rational point A of this presentation and note that $\|A\|^2 = \|A^*A\| = \max_\lambda |\lambda|$ where λ ranges over the eigenvalues of A^*A. The characteristic polynomial of a matrix can be found symbolically, and its roots can be found via the Argument Principle in complex analysis (see, for example, [10]). It is well known that the maximum function is a computable map from $\mathbb{R}^n$ to $\mathbb{R}$. The uniformity of these principles in n yields the uniform computability of the standard presentation of $M_n(\mathbb{C})$.

Later, we will use universal C*-algebras to give shorter and more sophisticated proof of this presentation's computability.

It now follows from the classification of finite-dimensional C*-algebras that every finite-dimensional C*-algebra is computably presentable. It also follows that certain

UHF algebras, such as the CAR algebra, are computably presentable. Recently, Eagle, Goldbring, McNicholl, and R. Miller announced the following in [11].

Theorem 7.10 *Suppose $\mathcal{A}$ is a UHF algebra. Then the following are equivalent.*

1. *$\mathcal{A}$ is computably presentable.*
2. *There is a nondecreasing sequence $(f_i)_{i\in\mathbb{N}}$ of computable functions from $\mathbb{N}$ into $\mathbb{N}$ so that $n(\mathcal{A}, p) = \lim_i f_i(p)$ for every prime p.*

7.2.3 Commutative C*-Algebras

By the Gelfand Theorem, every commutative unital C*-algebra is $*$-isomorphic to $C(X)$ for some compact Polish space X. It turns out that this principle holds effectively:

Theorem 7.11 *Suppose X is a compact Polish space. Then the following are equivalent.*

1. *X has a computably compact presentation.*
2. *$C(X)$ is computably presentable.*

Here, we must take some care with terminology. By a presentation of a *Polish space* X, we mean a triple $(X, d, (x_n)_{n\in\mathbb{N}})$ where d is a complete metric on X that is compatible with the topology on X and $(X, d, (x_n)_{n\in\mathbb{N}})$ is a presentation of (X, d). In other words, to define a presentation of a Polish space, one must define a metric *as well as* a sequence of distinguished points. Presentations of Polish spaces inherit the terminology applied to presentations of metric spaces. For example, a presentation $(X, d, (x_n)_{n\in\mathbb{N}})$ of a Polish space is computably compact if it is a computably compact presentation of (X, d).

The first direction of Theorem 7.11 was proved by Fox based on an observation of McNicholl [12]. This direction of the proof proceeds by assuming a computably compact presentation $X^\dagger = (X, d, (p_n)_{n\in\mathbb{N}})$ of X and then setting $f_n(p) = \frac{1}{1+d(p_n,p)}$ for all $p \in X$. It is easy to check that the f_n's separate points and so they generate $C^*(X)$ by the Stone-Weierstrass Theorem. It then follows from the computable compactness of $X^\dagger$ that the presentation $(C(X), (f_n)_{n\in\mathbb{N}})$ is computable. The second direction of Theorem 7.11 was proven in [8] and is considerably more involved. However, both directions of the proof of Theorem 7.11 are fully uniform.

As discussed in [13], there is a computably presentable compact Polish space with no computably compact presentation (see also [14], [15]). From this, we immediately obtain the following.

Corollary 7.12 *There is a computably presentable compact Polish space X so that $C(X)$ has no computable presentation.*

However, Fox also proved the following.

Theorem 7.13 *Let Z be a Stone space. Then Z is computably presentable if and only if $C(Z)$ is computably presentable.*

Let $\mathcal{A} = C(X)$. Suppose $\mathcal{A}^\dagger$ is a computable presentation of $\mathcal{A}$ and let $X^\dagger$ denote a computable presentation of X. It is then natural to wonder if the evaluation map $(f, x) \mapsto f(x)$ is a computable map from $C(X)^\dagger \times X^\dagger$ to $\mathbb{C}$. If it is, then we say that $C(X)^\dagger$ is *evaluative over* $X^\dagger$. It turns out that this condition characterizes the presentation of X and ensures its computable compactness. Moreover, the presentations built in the proof of Theorem 7.11 satisfy this relation. Building upon the methods used to prove Theorem 7.11, McNicholl [7] showed the following:

Theorem 7.14 *Suppose that X is a compact Polish space.*

1. *If $C(X)^\dagger$ is a computable presentation of $C(X)$, then, up to computable homeomorphism, there is a unique computably compact presentation $X^\dagger$ of X so that $C(X)^\dagger$ is evaluative over $X^\dagger$.*
2. *If $X^\dagger$ is a computable presentation of X over which a computable presentation of $C(X)$ is evaluative, then $X^\dagger$ is computably compact.*
3. *If $X^\dagger$ is a computably compact presentation of X, then, up to computable $*$-isomorphism, there is a unique computable presentation of $C(X)$ that is evaluative over $X^\dagger$.*

In his thesis [16], Fox showed that, for a subset X of the real line, $C(X)$ having a computable presentation implies an upper bound on the complexity of X.

Theorem 7.15 *If $X \subseteq \mathbb{R}$ is compact and $C(X)$ is computably presentable, then X is homeomorphic to a Δ^0_3-closed subset of $\mathbb{R}$.*

Fox proved this theorem by first establishing, for any compact X, that there is an embedding h of X into $\mathbb{R}$ that is $\Delta^0_3(\mathbf{d})$ in $C(X; \mathbb{R})$ where $\mathbf{d}$ is any Turing degree that computes the constant function $p \mapsto 1$ and the absolute value function. One can then show that $h[X]$ is a $\Delta^0_3(\mathbf{d})$-closed subset of $\mathbb{R}$. It remains to note that the constant function $p \mapsto 1$ and the absolute value function are computable functions of any computable presentation of $C(X)$, and thus the Turing degree $\mathbf{d}$ is not needed.

7.2.4 Presentations of Universal C-Algebras*

We now use the apparatus of universal C*-algebras to build presentations. To this end, we first give a more explicit construction of universal C*-algebras as in Blackadar [3]. It will also be useful to fix a standard sequence of distinct indeterminates. In fact, we just take our indeterminates to be the variables in the $*$-polynomials $x_0, x_1, \ldots$. Write $X_\omega = (x_j)_{j \in \mathbb{N}}$ and $X_n = (x_j)_{j \leq n}$. To simplify our exposition, we require our sequence of generators to be determined by our set of relations. So, when our set of relations $\mathcal{R}$ is infinite, we take X_ω to be our sequence of generators, while if $\mathcal{R}$ is finite, we assume our sequence of generators is X_n where n is the smallest number so that each variable that appears in $\mathcal{R}$ appears in X_n. Accordingly, we will suppress mention of $\mathcal{G}$ and just write $C^*\langle\mathcal{R}\rangle$.

We now summarize Blackadar's construction of the universal C*-algebra. Let $\mathfrak{A}^*_\omega$ denote the free $*$-algebra generated by X. Thus, $\mathfrak{A}^*_\omega$ is the set of all $*$-polynomials in the variables from X with no constant terms. Fix a set $\mathcal{R}$ of relations. When $p(x_0, \ldots, x_n)$ is a $*$-polynomial, set

$$\rho_{\mathcal{R}}(p(x_0, \ldots, x_n)) = \sup\{\|p(j(x_0), \ldots, j(x_n))\|_{\mathcal{A}} \,:\, (\mathcal{A}, j) \text{ is a representation of } \mathcal{R}\}.$$

As shown in Blackadar [3], if $\mathcal{R}$ is "reasonable," then $\rho_{\mathcal{R}}$ is a seminorm on $\mathfrak{A}^*_\omega$. Letting $\sim_{\mathcal{R}}$ denote the equivalence relation thereby induced on $\mathfrak{A}^*_\omega$, we may take $C^*\langle\mathcal{R}\rangle$ to be the completion of $\mathfrak{A}^*_\omega/\sim_{\mathcal{R}}$. We define the *standard presentation of* $C^*\langle\mathcal{R}\rangle$ to be the presentation whose j^{th} distinguished point is $[x_j]_{\sim_{\mathcal{R}}}$. We note that, due to our conditions on $\mathcal{R}$, the unit of a universal C*-algebra is a distinguished point of its standard presentation.

There is no reason to believe that the standard presentation of a universal C*-algebra is computable. However, we now describe a natural condition under which it is right c.e. A relation $\|p(x_1, \ldots, x_n)\| \leq a$ is called *rational* if p is a rational $*$-polynomial and a is a nonnegative dyadic rational.

Definition 7.16 Suppose $\mathcal{A}^\#$ is a presentation of a C*-algebra.

1. $\mathcal{A}^\#$ is *c.e.* if there is a c.e. set $\mathcal{R}$ of rational relations so that $\mathcal{A}^\#$ is computably $*$-isomorphic to $C^*\langle\mathcal{R}\rangle$.
2. $\mathcal{A}^\#$ is *finitely c.e.* if there is a finite set $\mathcal{R}$ of rational relations so that $\mathcal{A}^\#$ is computably $*$-isomorphic to $C^*\langle\mathcal{R}\rangle$.

Fox [12, Theorem 3.3] showed that c.e. presentations and right-c.e. presentations coincide. This principle has turned out to be extraordinarily useful for the development of the computability theory of C*-algebras. As a result, we shall dub it "Fox's Principle."

Theorem 7.17 (Fox's Principle) *Suppose $\mathcal{A}^\dagger$ is a presentation of a C*-algebra. Then $\mathcal{A}^\dagger$ is c.e. if and only if it is right c.e.*

Proof The forward direction of the proof is a fairly straightforward application of Pavelka-style completeness (Theorem 2.18). To see this, let T be the theory of C*-algebras (which is a c.e. set of sentences), and assume $\mathcal{A}^\dagger$ is the standard presentation of $C^*\langle\mathcal{R}\rangle$ where $\mathcal{R}$ is c.e. Each rational point of this presentation is a rational $*$-polynomial. We then observe that a rational number q belongs to the right Dedekind cut of $\|p(\bar{x})\|$ if and only if for some rational $q' < q$ there is a formal deduction of $\|p(\bar{x})\| \leq q'$ from $T \cup \mathcal{R}$. For the converse, we effectively enumerate all rational relations $\|p(\bar{x})\| \leq r$ that are true when x_j is replaced by the jth distinguished point of $\mathcal{A}^\dagger$. We note that the proof is uniform in both directions. □

We immediately obtain the following:

Corollary 7.18 *The standard presentations of the following* C^**-algebras are right c.e.:*

1. *The contraction algebras.*
2. *The Cuntz algebras.*
3. *The Toeplitz algebra.*
4. *The irrational rotation algebra* $\mathcal{A}_\theta$ *when* θ *is computable.*

Later, we will show that these algebras have computable presentations.

7.2.5 Presentations of Group C*-Algebras

Fox's Principle has an amazing range of applications. For example, the original proof of the following corollary, due to Fritz, Netzer, and Thom, used semidefinite programming [17, Corollary 2.2].

Corollary 7.19 *Suppose that* G *is a finitely presented group. Then the standard presentation of* $C^*(G)$ *is right c.e.*

Note that, by Theorem 7.17, the previous corollary holds under the weaker assumption that G is computably presented, that is, has a c.e. word problem.

It turns out that, for many finitely generated groups G, the standard presentation of the reduced group C^*-algebra $C^*_r(G)$ (that is, the presentation whose generators are the unitaries associated to the elements of G) is left c.e., as the following observation of Fritz, Netzer, and Thom [17, Lemma 3.1] shows.

Lemma 7.20 *If* G *is a finitely generated group with a solvable word problem, then the standard presentation of* $C^*_r(G)$ *is left c.e.*

The proof of the lemma follows from the equality

$$\|a^*a\|_r = \sup\{\tau(b^*a^*ab) \ : \ b \in \mathbb{C}[\Gamma], \ \tau(b^*b) \leq 1\},$$

where $\|\cdot\|_r$ denotes the norm on $C^*_r(\Gamma)$ and τ is the canonical trace on $\mathbb{C}[G]$ defined above.

Corollary 7.19 and Lemma 7.20 immediately imply the following.

Corollary 7.21 *If* G *is a finitely presented amenable group with a solvable word problem, then the standard presentation of* $C^*_r(G) = C^*(G)$ *is computable.*

7.2.6 Presentations of Simple C*-Algebras

Kuznetsov [18] proved that a computably presented simple group has a solvable word problem. Fox [12, Theorem 3.14] proved an analog for C*-algebras:

Theorem 7.22 *Every c.e. presentation of a simple* C*-*algebra is computable.*

Proof Suppose $\mathcal{A}^\dagger$ is a c.e. presentation of a simple C*-algebra. By Theorem 7.17, $A^\dagger$ is right c.e., whence it suffices to prove $\mathcal{A}^\dagger$ is left c.e.

Since $\mathcal{A}^\dagger$ is c.e., there is a c.e. set $\mathcal{R}$ of rational relations so that $\mathcal{A}^\dagger$ is the standard presentation of $\mathrm{C}^*\langle\mathcal{R}\rangle$. Let T denote the first-order theory of C*-algebras and expand the language of T so as to include a distinguished constant for each generator.

Fix a nonzero rational point ρ_0 of $A^\dagger$ and a positive dyadic number $l_0 < \|\rho_0\|_A$. Let ρ be any rational point of $\mathcal{A}^\dagger$, and let d be any positive dyadic number. The left-c.e.ness of $A^\dagger$ follows from the following claim: $\|\rho\| > d$ if and only if $T \cup \mathcal{R} \cup \{\|r(\bar{x})\| \le d\} \vdash \|q_0(\bar{x})\| \le l_0$. To see this, we first observe that, by the simplicity of $\mathcal{A}$, we have

$$\mathrm{C}^*\langle \mathcal{R} \cup \{\|\rho\| \le d\}\rangle = \begin{cases} \mathcal{A} & \text{if} \|\rho\| \le d \\ \{0\} & \text{otherwise.} \end{cases}$$

The claim then follows from Pavelka-style completeness. □

As the Cuntz algebras are simple (see, for example, [19]), we immediately obtain the following:

Corollary 7.23 *The standard presentations of the Cuntz algebras are computable.*

Since the irrational rotation algebras A_θ are simple (see, for example, [20]), we can also conclude the following:

Theorem 7.24 *For* $\theta \in (0, 1)$ *irrational, the following are equivalent.*

1. *θ is computable.*
2. *The standard presentation of A_θ is c.e.*
3. *The standard presentation of A_θ is computable.*

To see that (1) implies (2), we replace each relation in the definition of A_θ with its computable approximations. By the simplicity of A_θ (2) implies (3). Assume (3), and consider the identities

$$uvu^*v^* - vuv^*u^* = 2i\sin(2\pi\theta)\cdot \mathbf{1}_{\mathcal{A}} \text{ and } uvu^*v^* + vuv^*u^* = 2i\cos(2\pi\theta)\cdot \mathbf{1}_{\mathcal{A}}$$

(where u and v are the generators of A_θ). The left sides of these identities are computable vectors of the standard presentation of A_θ. It then follows that $\sin(2\pi\theta)$ and $\cos(2\pi\theta)$ are computable, whence so is θ.

Theorem 7.22 allows us to give a shorter proof that the standard presentation of $M_n(\mathbb{C})$ is computable. Indeed, since this presentation is finitely c.e. and $M_n(\mathbb{C})$ is simple, it follows from Theorem 7.22 that this presentation is computable.

The contraction algebras are not simple. Thus, to demonstrate the computability of their standard presentations, we need an additional tool, which we introduce now.

7.2.7 Residually Finite-Dimensional Presentations

A C*-algebra $\mathcal{A}$ is called *residually finite-dimensional (RFD)* if, for every nonzero $x \in \mathcal{A}$, there is a finite-dimensional representation π of $\mathcal{A}$ for which $\pi(x) \neq 0$. A group G is called residually finite-dimensional if $C^*(G)$ is a residually finite-dimensional C*-algebra.[4] Fritz, Netzer, and Thom [17, Theorem 1.5] showed that for RFD groups, the standard presentation of $C^*(G)$ is actually computable (instead of merely c.e.).

Theorem 7.25 *Suppose that G is a finitely presented residually finite-dimensional group. Then the standard presentation of* $C^*(G)$ *is computable.*

Examples of RFD groups include free groups and Fuschian groups [21]. By a theorem of Mal'cev [22], RFD groups are residually finite.[5] The converse implication does not hold. For example, $SL_3(\mathbb{Z})$ is residually finite but not RFD by a theorem of Bekka [23]. Even more drastically, Kirchberg [24] showed that the Connes Embedding Problem (see Sect. 8.4 below) is equivalent to the statement that the (residually finite) group $\mathbb{F}_2 \times \mathbb{F}_2$ is RFD. By the recent negative solution of the CEP [25], we now know that $\mathbb{F}_2 \times \mathbb{F}_2$ is not RFD.

Fox [12, Theorem 3.5] generalized Theorem 7.25 as follows:

Theorem 7.26 *If $\mathcal{A}$ is a finitely presented RFD* C*-*algebra, then any finitely c.e. presentation of $\mathcal{A}$ is computable.*

Proof Suppose that $\mathcal{A}^\dagger$ is finitely c.e. By Theorem 7.17, $\mathcal{A}^\dagger$ is right c.e., and so it suffices to show $\mathcal{A}^\dagger$ is left c.e. Choose a finite set of rational relations $\mathcal{R}$, and assume that $\mathcal{A}^\dagger$ is the standard presentation of $C^*\langle\mathcal{R}\rangle$. Thus, each rational point of $\mathcal{A}^\dagger$ has the form $q(\bar{x})$ for some rational $*$-polynomial q.

For each positive integer n, let $\mathbb{D}_n$ denote the closed unit ball of $\mathbb{C}^n$ and set

$$X_n = \{\bar{m} \in M_n(\mathbb{C}) \ : \ \mathcal{R}[\bar{m}] \text{ holds}\}.$$

Thus, X_n is a compact subset of the appropriate cartesian power of $\mathbb{R}$. Accordingly, for each rational point $q(\bar{x})$ of $\mathcal{A}^\dagger$, set

$$\alpha_{n,q} = \max\{\|q(\bar{m})\psi\| \ : \ \bar{m} \in X_n \ \wedge \ \psi \in \mathbb{D}_n\}.$$

The key claim of the proof is that $\|q(\bar{x})\|_{\mathcal{A}} = \sup_n \alpha_{n,q}^{1/2}$. To see this, we first note that since $\mathcal{A}$ is RFD, and since every finite-dimensional representation of $\mathcal{A}$ can be realized as a representation into a sufficiently large matrix algebra, $\|q(\bar{x})\|_{\mathcal{A}} \leq$

---

[4] This is not the definition of an RFD group given in [17], but is proven to be equivalent to the definition just given [17, Lemma 2.5].

[5] Recall that G is residually finite if, for every $x \in G \setminus \{e_G\}$, there is a homomorphism π of G into a finite group F so that $\pi(x) \neq e_F$.

$\sup_n \alpha_{n,q}^{1/2}$. On the other hand, the universality of $\mathcal{A}^\dagger$ guarantees that $\|q(\bar{x})\|_{\mathcal{A}} \geq \sup_n \alpha_{n,q}^{1/2}$.

Since each relation in $\mathcal{R}$ is rational, X_n is a computably compact set of $\mathbb{C}^n$ uniformly in n. Hence, $\alpha_{n,q}$ is a computable real uniformly in n and q. It now follows from the just-established identity that $\|q(\bar{x})\|_{\mathcal{A}}$ is left c.e. uniformly in q. Note that this proof is uniform in $\mathcal{R}$. □

Theorem 7.26 is a fairly significant generalization of Theorem 7.25. To illustrate this point, we use Theorem 7.26 to show that many other C*-algebras have computable standard presentations. For example, since the universal contraction algebras $\mathcal{F}(n)$ and $\mathcal{F}(\omega)$ are RFD [26], we obtain the following.

Corollary 7.27 *For any $n \in \mathbb{N} \cup \{\omega\}$, the standard presentation of $\mathcal{F}(n)$ is computable.*

We note that the case of $\mathcal{F}(n)$ for $n \in \mathbb{N}$ is immediately implied by Theorem 7.26. The case of $\mathcal{F}(\omega)$ follows from the uniform computability of the presentation of $\mathcal{F}(n)$.

Although the Toeplitz algebra $\mathcal{T}$ is neither simple nor RFD, we can show that its standard presentation is computable by means of a more "bare hands" approach.

Theorem 7.28 *The standard presentation of the Toeplitz algebra is computable.*

Indeed, since the standard presentation of the Toeplitz algebra is finitely c.e., it suffices to show it is left c.e. As mentioned above, an alternate description of $\mathcal{T}$ is as the C*-algebra generated by the right shift operator S on ℓ^2. Every monomial in S and S^* can (effectively) be put in the form $S^m(S^*)^n$ for some $m, n \in \mathbb{N}$. The action of such a monomial is easily described on the basis vectors of ℓ^2. It is then a straightforward exercise to show that the standard presentation of $\mathcal{T}$ is left c.e.

7.2.8 Word Problems

A classical concern in group theory is understanding which groups have a solvable word problem. Here, we discuss the analogous issue in the C*-algebra context. All of the results in this subsubsection are due to Fox [12, Sect. 3].

First, we must define what we mean by the word problem for a presentation $\mathcal{A}^\dagger$ of a C*-algebra. Let s_n denote the nth distinguished point of $\mathcal{A}^\dagger$. By the universal property of the universal contraction algebras, there is a canonical surjective $*$-homomorphism $\pi_{A^\dagger}$ of $\mathcal{F}(\omega)$ onto $\mathcal{A}$, and this map sends the nth generator of $F(\omega)$ to $(\|s_n\| + 1)^{-1} s_n$. This observation leads to the following.

Definition 7.29 Let $\mathcal{A}$ be a C*-algebra, and suppose $\mathcal{A}^\dagger$ is a presentation of $\mathcal{A}$. The *word problem* of $\mathcal{A}^\dagger$, denoted WP($\mathcal{A}^\dagger$), is the kernel of $\pi_{\mathcal{A}^\dagger}$.

Fox connected the computability-theoretic properties of $\mathcal{A}^\dagger$ and the computability-theoretic properties of the closed set $\mathrm{WP}(\mathcal{A}^\dagger)$. For example, the following theorem is [12, Theorem 3.11].

Theorem 7.30 *Let $\mathcal{A}$ be a* C**-algebra with presentation $\mathcal{A}^\dagger$. Then $\mathcal{A}^\dagger$ is c.e. if and only if* $\mathrm{WP}(\mathcal{A}^\dagger)$ *is a c.e. closed set of $F(\omega)$.*

The idea behind the proof of the previous theorem is that for a rational point ρ of $\mathcal{A}^\dagger$ and a rational number q, $B(\rho, q) \cap \mathrm{WP}(\mathcal{A}^\dagger) \neq \emptyset$ if and only if $\|\rho\| < q$.

By a standard Craig's Trick argument, from a c.e. presentation $\mathcal{A}^\dagger$ of $\mathcal{A}$, it is possible to form a presentation $\mathcal{A}^\#$ of $\mathcal{A}$ with a computable set of relations. This observation leads us to the following.

Theorem 7.31 (Fox) *Let $\mathcal{A}$ be a* C**-algebra with presentation $\mathcal{A}^\dagger$. Then $\mathcal{A}^\dagger$ is computable if and only if the word problem of $\mathcal{A}^\dagger$ is strongly computably closed.*

As remarked in Chap. 3, there are c.e. closed sets of $\mathbb{R}$ that are not strongly computably closed (see Example 3.39). This leads to the following.

Question 7.32 Is there a presentation $\mathcal{A}^\dagger$ of a C*-algebra $\mathcal{A}$ such that $\mathrm{WP}(\mathcal{A}^\dagger)$ is computably closed and yet $\mathcal{A}^\dagger$ is not computable?

We now specialize the discussion to the case of group C*-algebras. We first mention two results of Fritz, Netzer, and Thom [17, Remark 1.2 and Theorem 1.3].

Lemma 7.33 *Suppose that G is a finitely generated group such that the standard presentation of either $\mathrm{C}^*(G)$ or $\mathrm{C}^*_r(G)$ is computable. Then G has a solvable word problem.*

The proof of the lemma is quite simple. For $g \in G$, one has $\|u_g - u_e\| = 0$ if $g = e$. However, if $g \neq e$, then we have $\|u_g - u_e\| \geq \|u_g - u_e\|_r \geq \|(u_g - u_e)\delta_e\| = \|\delta_g - \delta_e\| = \sqrt{2}$.

Corollary 7.34 *If G is a finitely presented amenable group, then the standard presentation of $\mathrm{C}^*(G) = \mathrm{C}^*_r(G)$ is computable if and only if G has a solvable word problem.*

Fox gave a much more thorough analysis of the connection between group presentations and the associated presentations of the corresponding group C*-algebras. Before explaining these results, let us first discuss presentations of groups.

Suppose that G is a (countable) group. By a *presentation* of G, we mean a pair $(G, \bar{g})$ such that $\bar{g}$ generates G. Of course, this is the same as a presentation of G viewed as a structure in discrete logic (equivalently, as a structure in continuous logic when equipped with the discrete metric). Note that a c.e. (respectively finitely c.e.) presentation of a group is what a group theorist would call a computable (respectively finite) presentation of the group.

If $G^\dagger$ is a presentation of G, then one obtains induced presentations $\mathrm{C}^*(G^\dagger)$ and $\mathrm{C}^*_r(G^\dagger)$ of $\mathrm{C}^*(G)$ and $\mathrm{C}^*_r(G)$, respectively, by defining the distinguished points to be the unitaries corresponding to the generators of G. It is readily verified that if $G^\dagger$ is (finitely) c.e., then so is the presentation $C^*(G^\dagger)$. The following theorem is [12, Theorem 3.20].

Theorem 7.35 *Let G be a countable discrete group with presentation $G^\dagger$. Then the following are equivalent.*

1. *$G^\dagger$ is a computable presentation of G.*
2. $\mathrm{WP}(C^*(G^\dagger))$ *is c.e. closed, and* $\mathrm{WP}(C^*_r(G^\dagger))$ *is strongly co-c.e. closed.*
3. $\mathrm{WP}(C^*(G^\dagger))$ *is c.e. closed, and the set of rational points of $F(\omega)$ that belong to the complement of* $\mathrm{WP}(C^*(G^\dagger))$ *is c.e.*
4. $\mathrm{WP}(C^*_r(G^\dagger))$ *is co-c.e. closed, and there is a c.e. set of open rational balls, each of which intersects* $\mathrm{WP}(C^*_r(G^\dagger))$*, such that the set contains all open balls centered at rational points belonging to* $\mathrm{WP}(C^*_r(G^\dagger))$.

Corollary 7.36 *If either* $\mathrm{WP}(C^*(G^\dagger))$ *or* $\mathrm{WP}(C^*_r(G^\dagger))$ *is computably closed, then $G^\dagger$ is a computable presentation of G.*

Question 7.37 Is the converse of the previous corollary true? In other words, if $G^\dagger$ is computable, must $\mathrm{WP}(C^*(G^\dagger))$ and $\mathrm{WP}(C^*_r(G^\dagger))$ be computably closed?

Note that a finitely c.e. presentation $G^\dagger$ is computable precisely when G has a solvable word problem. Thus, the following corollary of Theorem 7.35 generalizes Lemma 7.33.

Corollary 7.38 *If G is amenable, then $G^\dagger$ is computable if and only if* $C^*(G^\dagger) = C^*_r(G^\dagger)$ *is computable.*

7.2.9 *Locally Universal C*-Algebras and the Kirchberg Embedding Problem*

Computability theory played a significant role in the recent resolution of the Connes Embedding Problem (discussed in Sect. 8.4 below). We now make the case that computability theory could be useful for resolving another conjecture in operator algebras, namely the Kirchberg Embedding Problem. The results of this section are due to Fox, Goldbring, and Hart [27].

To define this problem, we first define a C*-algebra $\mathcal{A}$ to be *locally universal* if every C*-algebra embeds into an ultrapower of $\mathcal{A}$. It is a basic fact of model theory that locally universal C*-algebras exist. Note also that any C*-algebra containing a locally universal C*-algebra is itself locally universal. The *Kirchberg Embedding Problem (KEP)*, a natural C*-algebra version of the Connes Embedding Problem, asks whether or not the Cuntz algebra $\mathcal{O}_2$ is locally universal. For more information on the KEP and locally universal C*-algebras, including why $\mathcal{O}_2$ is the only possible Cuntz algebra that could be locally universal, the reader may consult [28].

Our goal now is to show that a positive solution of the KEP implies a certain very strong, and, we believe, implausible, statement in computability theory. In order to formulate this statement, we need the concept of computable weak stability. Informally, a set $\mathcal{R}$ of relations is weakly stable if it is the case that whenever a tuple $\bar{a}$ in a C*-algebra $\mathcal{A}$ "comes close to satisfying $\mathcal{R}$," then $\bar{a}$ is close to a tuple $\bar{b}$ in $\mathcal{A}$ that *does* satisfy $\mathcal{R}$. Computable weak stability requires that how nearly $\bar{a}$ must satisfy $\mathcal{R}$ can be computed from how close we desire $\bar{a}$ and $\bar{b}$ to be. The following definition formalizes these intuitions.

Definition 7.39 Suppose that $\mathbf{x} = (x_1, \ldots, x_N)$ is a tuple of variables, $C_1, \ldots, C_N \in \mathbb{N}$, and $p_1(\mathbf{x}), \ldots, p_M(\mathbf{x})$ are rational $*$-polynomials. Consider the set

$$\mathcal{R} = \{p_i(\mathbf{x}) = 0 \ : \ i = 1, \ldots, M\} \cup \{\|x_j\| \leq C_j \ : \ j = 1, \ldots, N\}$$

of relations.

1. For a C*-algebra $\mathcal{B}$ and $w_1, \ldots, w_N \in \mathcal{B}$, let $\mathcal{R}^{\mathcal{B}}$ denote the quantity

$$\max(\{\|p_i(\mathbf{w})\| \ : \ i \in \{1, \ldots, M\}\} \cup \{\|w_j\| - C_j \ : j \in \{1, \ldots, N\}\})$$

   and write $\mathcal{B} \models \mathcal{R}(\mathbf{w})$ if $p_i(\mathbf{w}) = 0$ for all $i \in \{1, \ldots, M\}$ and $\|w_j\| \leq C_j$ for all $j \in \{1, \ldots, N\}$.
2. A function $g : \mathbb{N} \to \mathbb{N}$ is a *modulus of weak stability* for $\mathcal{R}$ provided that, for every $k \in \mathbb{N}$, every C*-algebra $\mathcal{B}$, and every $w_1, \ldots, w_N \in \mathcal{B}$, if $\mathcal{R}^{\mathcal{B}}(\mathbf{w}) < 2^{-g(k)}$, then there exists $z_1, \ldots, z_N \in \mathcal{B}$ so that $\max_j \|w_j - z_j\| < 2^{-k}$ and $\mathcal{B} \models \mathcal{R}(\mathbf{z})$.
3. We say that $\mathcal{R}$ is *weakly stable* if it has a modulus of weak stability and *computably weakly stable* if it has a computable modulus of weak stability.

Definition 7.40 Let $\mathcal{A}$ be a C*-algebra.

1. $\mathcal{A}$ is *weakly stable* if there is a weakly stable $\mathcal{R}$ so that $\mathcal{A}$ is $*$-isomorphic to C$^*\langle\mathcal{R}\rangle$.
2. A presentation of $\mathcal{A}$ is *computably weakly stable* if it is computably $*$-isomorphic to C$^*\langle\mathcal{R}\rangle$ for some computably weakly stable $\mathcal{R}$.

Examples 7.41 The following are all weakly stable C*-algebras.

1. C$^*(\mathbb{F}_n)$ for $n \geq 1$.
2. Any finite-dimensional C*-algebra.
3. The Cuntz algebras $\mathcal{O}_n$.
4. The Toeplitz algebra $\mathcal{T}$.

For proofs of the statements in the previous example, see [29]. The next results indicate the robustness of the notion of being computably weakly stable.

Theorem 7.42 *Suppose $\mathcal{R}$ and $\mathcal{S}$ are finite sets of relations, and assume $\mathcal{R}$ is computably weakly stable. If* C$^*\langle\mathcal{R}\rangle$ *is computably $*$-isomorphic to* C$^*\langle\mathcal{S}\rangle$*, then $\mathcal{S}$ is computably weakly stable.*

Theorem 7.43

1. *The direct sum of two computably weakly stable presentations of* C*-*algebras is computably weakly stable.*
2. *The standard presentations of the* C*-*algebras in Example 7.41 are computably weakly stable.*

The mere existence of a locally universal C*-algebra with a computable presentation has a very surprising implication.

Theorem 7.44 *If there exists a locally universal* C*-*algebra with a computable presentation, then every computably weakly stable presentation of a* C*-*algebra is computable.*

Proof Suppose that $\mathcal{A}^\dagger$ is a computably weakly stable presentation of $\mathcal{A}$ and that $\mathcal{B}^\dagger$ is a computable presentation of a locally universal C*-algebra $\mathcal{B}$. Since $\mathcal{A}^\dagger$ is right c.e., it suffices to show that it is also left c.e. Given a rational point $q(\bar{x})$ of $\mathcal{A}^\dagger$, we seek an increasing computable sequence of lower bounds to $\|q(\bar{x})\|$. For each $j \in \mathbb{N}$, first determine $n \in \mathbb{N}$ such that 2^{-n} is a witness to the uniform continuity of each of the polynomials p_i corresponding to 2^{-j}, and then let $m \in \mathbb{N}$ be as in the definition of computably weakly stable for this choice of n. Now enumerate over all rational points $\bar{r}$ of $\mathcal{B}^\dagger$ such that $\|p_i(\bar{r})\| < 2^{-m}$ for all i and such that $\|r_k\| < C_k + 2^{-m}$. Now enumerate over all dyadic rationals d and output d whenever d is greater than all previous outputs of the algorithm and $\|q(\bar{r})\| > d + 2^{-j}$; this is possible since $\mathcal{B}^\dagger$ is computable. Note then that d is a lower bound to $\|q(\bar{x})\|$. Indeed, by the choice of m, there is a $*$-homomorphism $\varphi : \mathcal{A} \to \mathcal{B}$ such that $\|\varphi(x_k) - r_k\| < 2^{-n}$ for all k. Consequently, we have

$$\|q(\bar{x})\| \geq \|\varphi(q(\bar{x}))\| = \|q(\varphi(\bar{x}))\| \geq \|q(\bar{r})\| - 2^{-j} > d.$$

We claim that the above sequence does indeed converge to $\|q(\bar{x})\|$. To see this, take $d < \|q(\bar{x})\|$ and $j \in \mathbb{N}$ such that $\|q(x)\| > d + 2^{-j}$. Let n and m be as above. Since $\mathcal{B}$ is locally universal, $\mathcal{A}$ embeds into $\mathcal{B}^{\mathcal{U}}$. Thus there is, for sufficiently small $\epsilon > 0$, $\bar{z} \in B$ such that $\|p_i(\bar{z})\| < 2^{-m} - \epsilon$ for each i and $\|z_k\| < C_k + 2^{-m} - \epsilon$ for all k and $\|q(\bar{z})\| > d + 2^{-j} + \epsilon$. Then there is a rational point $\bar{r}$ of B such that $\|p_i(\bar{r})\| < 2^{-m}$ for all i and $\|r_k\| < C_k + 2^{-m}$ for all k and $\|q(\bar{r})\| > d + 2^{-j}$. Consequently, in the previous paragraph, d would have been enumerated. □

Since O_2 has a computable presentation, we see that a positive solution to the KEP would imply that every computably weakly stable presentation is computable. This implication may offer a path to refuting the KEP.

7.2.10 Presentations of von Neumann Algebras

As is the case for C*-algebras, tracial von Neumann algebras may be viewed as metric structures in an appropriate language. Once again, the sorts in the language are intended to capture the operator norm balls of the algebra. However, unlike the case of C*-algebras, the metric on the sorts is not that induced by the operator norm. One motivation for not using the operator norm is that the corresponding ultraproduct construction would then be the usual C*-algebraic ultraproduct. However, the C*-algebraic ultraproduct of von Neumann algebras is rarely a von Neumann algebra again. Instead, the metric on each sort is that induced by the norm $\| \cdot \|_\tau$. (Of course, one must check that the algebra operations are uniformly continuous with respect to this metric on each sort and that the sorts are complete with respect to this metric.) When tracial von Neumann algebras are viewed this way, then the model-theoretic ultraproduct coincides with the *tracial ultraproduct* construction already considered by operator algebraists [6]. It is important to point out that the language for tracial von Neumann algebras does not have a predicate symbol for the operator norm. The only role the operator norm plays in the language is in determining the sorts. We also point out that the zero vector and the unit are named by constant symbols in the language.

As before, we now have natural notions of presentations of tracial von Neumann algebras. It is important to stress now that a presentation of a tracial von Neumann algebra is a countable subset of the algebra that generates a $\| \cdot \|_\tau$-dense subset, or, equivalently, an SOT-dense subset. We note that the computability of a presentation of a von Neumann algebra does not imply the computability of the operator norm on the rational points, but it does imply the computability of τ on these points.

Example 7.45 An example of a II_1 factor that admits a computable presentation is the hyperfinite II_1 factor $\mathcal{R}$. One way to see this is to use an alternate definition of $\mathcal{R}$ (in fact, the definition of $\mathcal{R}$ offered by Murray and von Neumann in [30]). We first note that the canonical embedding of $M_{2^n}(\mathbb{C})$ into $M_{2^{n+1}}(\mathbb{C})$ preserves the trace. Thus, the inductive limit $\mathcal{A} = \lim_n M_{2^n}(\mathbb{C})$ is naturally a $*$-algebra with a "tracial state" τ. By employing an analog of the GNS construction, we have a representation $\pi_\tau : \mathcal{A} \to B(\mathcal{H})$. The von Neumann algebra $\pi_\tau(\mathcal{A})'' \subseteq \mathcal{B}(\mathcal{H})$ is isomorphic to $\mathcal{R}$. Moreover, the unique trace on $\mathcal{R}$ corresponds to the trace on $\pi(\mathcal{A})''$ given by the vector state $\langle \cdot \hat{1}, \cdot \rangle_\tau$ where the unit of $\mathcal{A}$ corresponds to the identity matrix in the various $M_{2^n}(\mathbb{C})$'s.

We obtain a computable presentation of $\mathcal{R}$ by letting the distinguished points be the matrices in the various $M_{2^n}(\mathbb{C})$'s with entries in $\mathbb{Q}(i)$; note that the rational points of this presentation are in fact all special and that the trace is indeed computable on such points.

Example 7.46 We now consider the case of the group von Neumann algebra $L(G)$ associated to a countable discrete group G. Consider some fixed enumeration $(g_n)_{n\in\mathbb{N}}$ of elements that generate G, and consider the presentation $L(G)^\dagger$ of $L(G)$ whose distinguished points are the associated canonical unitaries $(u_{g_n})_{n\in\mathbb{N}}$. Given that the trace of an element of $\mathbb{C}[G]$ is simply the coefficient of u_e, it is straightforward to verify that the presentation $L(G)^\dagger$ is computable precisely when the set of words $w(g_1, \ldots, g_k)$ that evaluate to the identity is computable. In particular, if G is a finitely generated group with a solvable word problem, then the associated presentation of $L(G)$ is computable.

Combined with Connes' result mentioned above, we see that any infinite ICC amenable group with a solvable word problem yields a computable presentation of $\mathcal{R}$.

7.2.11 Subfactors

A *subfactor* is an inclusion $N \subseteq M$ of factors. Given a subfactor $N \subseteq M$ of a tracial factor, one has an inclusion of Hilbert spaces $L^2(N) \subseteq L^2(M)$, whence there is an orthogonal projection map $e : L^2(M) \to L^2(N)$. It is a basic fact in the subject that $e(M) \subseteq N$. The map $E = e|M : M \to N$ is called the *conditional expectation of* M *onto* N. The conditional expectation map is central to the study of subfactor theory. Hence, its computability is a natural topic for investigation. The first work along these lines was undertaken by Goldbring in [31], which we now explain.

In the context of the previous paragraph, we call a pair of presentations $(M^\dagger, N^\#)$ *computable* if $M^\dagger$ is computable and the inclusion map is a computable map from $N^\#$ into $M^\dagger$. We note that the latter condition implies the computability of $N^\#$.

The proof of the following lemma is straightforward.

Lemma 7.47 *Suppose that* $(M^\dagger, N^\#)$ *is a computable pair. Then the following are equivalent.*

1. *The conditional expectation of* M *onto* N *is a computable map from* $M^\dagger$ *to* $N^\#$.
2. *The map* $p \mapsto d(p, N)$ *is a computable map from* $M^\dagger$ *to* $\mathbb{R}$.

We note that the latter condition is equivalent to saying that N is a computably closed set of $M^\dagger$.

Goldbring's main result gave natural conditions under which the conditional expectation map for a computable pair is computable. The first of these conditions is essentially model theoretic.

Definition 7.48 M is an *existentially closed (e.c.) tracial von Neumann algebra* if whenever $M \subseteq N$ is an inclusion of tracial von Neumann algebras, there is an embedding of N into an ultrapower $M^{\mathcal{U}}$ of M whose restriction to M is the diagonal embedding.

This semantic definition is equivalent to the syntactic definition, namely $\sigma^M = \sigma^N$ for all extensions N of M and all existential formulas with parameters from M. Existentially closed tracial von Neumann algebras are automatically II_1 factors, whence one also sees the term e.c. factor in the literature. For more on e.c. factors, one may consult [32]. Clearly, the notion of an existentially closed tracial von Neumann algebra makes sense for an arbitrary model of an arbitrary theory, and we will return to this notion in Sect. 8.3 below.

The next condition in Goldbring's result involves the notion of a Kazhdan presentation, which is based on the classical notions of Kazhdan sets and pairs. We define these as follows.

Definition 7.49 Let N be a factor. Let F be a finite subset of the unit ball of N, and let $K > 0$.

1. We say that (K, F) is a *Kazhdan pair for* N if, for all sufficiently small $\delta > 0$, all tracial von Neumann algebras M containing N and all contractions $x \in M$ so that $\|xy - yx\|_\tau < \delta$ for all $y \in F$, there is a contraction $z \in M$ such that $\|x - z\|_\tau < K\delta$ and $yz = zy$ for all $y \in N$.
2. We say that F is a *Kazhdan set for* N if (K', F) is a Kazhdan pair for some $K' > 0$.
3. A presentation $N^\#$ is a *Kazhdan presentation* if the set of its rational points includes a Kazhdan set for N.

The final condition in Goldbring's theorem is purely classical and is known as "property (T)." The notion of property (T) for a II_1 factor N is defined using so-called bimodules over N. Examples of property (T) II_1 factors include $L(G)$ for G a property (T) group such as $G = \mathrm{SL}_3(\mathbb{Z})$. While we do not give a definition here, we mention one important consequence of property (T): if N is a property (T) II_1 factor, then it has a Kazhdan set.

We can now state Goldbring's result:

Theorem 7.50 *Let $N \subseteq M$ be an inclusion of subfactors that satisfy the following:*

1. *M is an existentially closed tracial von Neumann algebra.*
2. *N is a property (T) subfactor.*

If $(M^\dagger, N^\#)$ is a computable pair, and if $N^\#$ is a Kazhdan presentation, then the conditional expectation of M onto N is a computable map from $M^\dagger$ to $N^\#$.

We note that Theorem 7.50 relativizes in the obvious way.

In [33], Jones defined the *index* of a subfactor $N \subseteq M$, denoted $[M : N]$. This index is an element of $[1, \infty]$ and need not be an integer. The motivation behind the terminology is the result that for a subgroup H of a group G, one has that $[L(G) : L(H)] = [G : H]$. The subfactors $N \subseteq M$ considered in the previous theorem are of infinite index. A study of computability for finite index subfactors was undertaken by Fox and Goldbring in [34]. There they connected the computability of the conditional expectation map with two other important concepts in subfactor theory: the *Jones basic construction* and *Pimsner-Popa bases*. We explain these forthwith.

Suppose that $N \subseteq M$ is a subfactor of a II_1 factor, and let $e : L^2(M) \to L^2(M)$ denote the orthogonal projection map. Viewing $M \subseteq \mathcal{B}(L^2(M))$, we define the *Jones basic construction* to be the von Neumann algebra $M_1 = (M \cup \{e\})'' \subseteq \mathcal{B}(L^2(M))$. M_1 is a II_1 factor if and only if $[M : N] < \infty$. If $[M : N]$ is finite, then $[M_1 : M] = [M : N]$, and the trace on M_1 is determined by the fact that $\tau(e) = [M : N]^{-1}$. If $M^\#$ is a presentation of M, the presentation of $M_1^\#$ given by adding e to the distinguished points of $M^\#$ is called the *induced presentation on* $M_1^\#$.

Suppose that $[M : N] < \infty$ has integer part n. Then there exist elements $m_1, \ldots, m_{n+1}$ of M with the following properties:

- $E_N(m_j^* m_k) = 0$ for all distinct $j, k \in \{1, \ldots, n+1\}$.
- $E_N(m_j^* m_j) = 1$ for all $j \in \{1, \ldots, n\}$.
- $E_N(m_{n+1}^* m_{n+1})$ is a projection of trace $[M : N] - n$.

(By convention, $m_{n+1} = 0$ if $[M : N] \in \mathbb{N}$.) Any such family of elements is called a *Pimsner-Popa basis of M over N*. The terminology is inspired by the fact that if $m_1, \ldots, m_{n+1}$ is a Pimsner-Popa basis of M over N, then every $x \in M$ can be written uniquely as $x = \sum_{j=1}^{n+1} m_j x_j$ with each $x_j \in N$ for $j = 1, \ldots, n$ and $x_{n+1} \in E_N(m_{n+1}^* m_{n+1})N$. Other important properties of a Pimsner-Popa basis are the following.

- $\sum_{j=1}^{n+1} m_j e_N m_j^* = 1$.
- $\sum_{j=1}^{n} m_j m_j^* = [M : N]$.

If $M^\#$ is a presentation of M, then a Pismner-Popa basis $m_1, \ldots, m_{n+1}$ is a *computable Pimsner-Popa basis of $M^\#$ over N* if it satisfies the following:

1. Each m_j is a computable point of $M^\#$.
2. Each $E_N(m_j)$ is a computable point of $M^\#$.
3. $E_N(m_{n+1}^* m_{n+1})$ is a computable point of $M^\#$.

The following is the main result of Fox and Goldbring [34].

Theorem 7.51 *Suppose that N is a subfactor of M such that $[M : N] < \infty$. Fix a computable presentation $M^\#$ of M, and let $M_1^\#$ be the induced presentation of the Jones basic construction M_1. Then the following are equivalent.*

1. *$M_1^\#$ is computable.*
2. *The following two statements hold:*
   a. *$E_N : M^\# \to M^\#$ is computable.*
   b. *$[M : N]$ is a computable real number.*
3. *The following two statements hold:*
   a. *$M^\#$ is computable, and there is a presentation $N^\dagger$ for which the inclusion map $N^\dagger \hookrightarrow M^\#$ is computable.*
   b. *There is a computable Pimsner-Popa basis of $M^\#$ over N.*

Once again, Theorem 7.51 relativizes.

An amazing theorem of Jones [33] is that the indices of finite-index subfactor pairs are precisely the elements of $\{4\cos^2(\frac{\pi}{n}) : n = 3, 4, 5, \ldots\} \cup [4, \infty]$. Moreover, all of these indices are realized by subfactors of $\mathcal{R}$. We see that in the case that $[M : N] = 4\cos^2(\frac{\pi}{n})$, which is a computable number, item 2b in the previous theorem automatically holds.

Examples of subfactors which satisfy the equivalent conditions of the previous theorem are presented in [34].

7.3 Computable Categoricity of Operator Algebras

The results in this section concern the computability of $*$-homomophisms between C*-algebras. We begin with a very useful characterization of the computability of these maps, which follows immediately from Lemma 3.49 and the fact that $*$-homomorphisms between C*-algebras are 1-Lipschitz.

Proposition 7.52 *Suppose $\mathcal{A}^\#$ and $\mathcal{B}^\#$ are presentations of* C**-algebras, and let φ be a $*$-homomorphism from $\mathcal{A}$ into $\mathcal{B}$. Then φ is a computable map from $\mathcal{A}^\#$ to $\mathcal{B}^\#$ if and only if it is computable on the rational points of $\mathcal{A}^\#$.*

Proposition 7.52 further demonstrates the power of the C*-identity. For example, there are incomputable uniformly continuous functions $f : \mathbb{R} \to \mathbb{R}$ that are nevertheless computable on the rational numbers (see, for example, [35, Theorem 6]).

In the finite-dimensional case, we have the following result of Fox [12, Corollary 5.9].

Theorem 7.53 *Finite-dimensional* C**-algebras are computably categorical.*

Although the statement of the previous theorem is perhaps unsurprising, Fox's proof is surprisingly involved. In particular, Fox's proof proceeds by carefully constructing a set of matrix units for a finite-dimensional C*-algebra $\mathcal{A}$ from one of its computable presentations. Using the technology of computable weak stability, it is now possible to give a shorter proof by means of a "brute force" search procedure for these units.

Theorem 7.14 can be used to classify the computably categorical C*-algebras that are unital and commutative [7].

Theorem 7.54 *Suppose X is a compact Polish space. Then the following are equivalent.*

1. *$C(X)$ is computably categorical.*
2. *Any two computably compact presentations of X are computably homeomorphic.*

Fox showed that even very basic abelian C*-algebras can fail to be computably categorical [12, Corollary 4.7]:

Theorem 7.55 *$C([0, 1])$ is not computably categorical.*

The proof of Theorem 7.55 proceeds by showing that an abelian C*-algebra $\mathcal{A}$ is computably categorical if and only if the set of self-adjoint elements $\mathcal{A}_{\mathrm{sa}}$ of $\mathcal{A}$ is computably categorical as a *real* C*-algebra. This, in turn, is equivalent to $\mathcal{A}_{\mathrm{sa}}$ being computably categorical as a real Banach algebra. In the case of $C([0, 1])$, its set of self-adjoint elements is the set $C([0, 1]; \mathbb{R})$ of real-valued continuous functions on $[0, 1]$. As discussed in Chap. 5, $C([0, 1]; \mathbb{R})$ was shown not to be computably categorical as a real Banach algebra by Melnikov and Ng [36]. An immediate corollary of the previous theorem is the following.

Corollary 7.56 *There exist computably compact presentations of* $[0, 1]$ *that are not computably homeomorphic.*

By means of Theorem 7.54 and the effective Stone duality (see [37]), we obtain the following.

Corollary 7.57 $C(2^\omega)$ *is computably categorical.*

This contrasts with the aforementioned result of Thewmorokat that $C(2^\omega; \mathbb{R})$ is not computably categorical as a Banach space.

In his thesis [16], Fox established some positive results about relative categoricity for certain abelian C*-algebras.

Theorem 7.58 *Suppose that* X *is a compact subset of* $\mathbb{R}$.

1. *If* X *is the union of finitely many points and closed intervals, then* $C(X)$ *is* Δ^0_3-*categorical.*
2. *If* X *is the union of finitely many points and closed intervals together with a nonempty perfect set, then* $C(X)$ *is* Δ^0_5-*categorical.*

Again, via Theorem 7.54, we obtain the following:

Corollary 7.59 *Suppose that* X *is a compact subset of* $\mathbb{R}$.

1. *If* X *is the union of finitely many points and closed intervals, then any two computably compact presentations of* X *are* Δ^0_3-*homeomorphic.*
2. *If* X *is the union of finitely many points and closed intervals together with a nonempty perfect set, then any two computably compact presentations of* X *are* Δ^0_5-*homeomorphic.*

Leaving the realm of abelian algebras, Eagle, Goldbring, McNicholl, and R. Miller recently announced the following in [11].

Theorem 7.60 *Every UHF algebra is computably categorical.*

The proof uses computable weak stability to search for systems of matrix units and thereby build an explicit inductive limit representation for each computably presented UHF algebra. Constructing $*$-isomorphisms between those of the same type is a fairly straightfoward back-and-forth construction.

Computable categoricity in the context of tracial von Neumann algebras seems not to have been investigated thus far.

Question 7.61 Are any of the computably presentable tracial von Neumann algebras discussed here computably categorical?

References

1. Isaac Goldbring. *Model Theory of Operator Algebras*, volume 11. Walter de Gruyter GmbH & Co KG, 2023.
2. M. Rørdam, F. Larsen, and N. Laustsen. An introduction to K-theory for C*-algebras, volume 49 of *London Mathematical Society Student Texts*. Cambridge University Press, Cambridge, 2000.
3. Bruce Blackadar. Shape theory for C*-algebras. *Math. Scand.*, 56(2):249–275, 1985.
4. Lewis A. Coburn. The C*-algebra generated by an isometry. *Bull. Amer. Math. Soc.*, 73:722–726, 1967.
5. Alain Connes. Classification of injective factors: Cases II_1, II_∞, III_λ, $\lambda \neq 1$. *Annals of Mathematics*, 104(1):73–115, 1976.
6. Ilijas Farah, Bradd Hart, and David Sherman. Model theory of operator algebras II: Model theory. *Israel Journal of Mathematics*, 201:477–504, 2010.
7. Timothy H. McNicholl. Evaluative presentations. *J. Logic Comput.*, 35(5):Paper No. exaf036, 11, 2025.
8. Peter Burton, Christopher J. Eagle, Alec Fox, Isaac Goldbring, Matthew Harrison-Trainor, Timothy H. McNicholl, Alexander Melnikov, and Teerawat Thewmorakot. Computable Gelfand duality. *Proc. Amer. Math. Soc.,* to appear.
9. Isaac Goldbring. Computably strongly self-absorbing C*-algebras. *Theoretical Computer Science*, to appear.
10. A. Matheson and T.H. McNicholl. Computable analysis and Blaschke products. *Proceedings of the American Mathematical Society*, 136(1):321–332, 2008.
11. Chris Eagle, Isaac Goldbring, Timothy McNicholl, and Russell Miller. Computable K-theory for C*-algebras: UHF algebras. *Trans. Amer. Math. Soc.,* to appear.
12. Alec Fox. Computable presentations of C*-algebras. *J. Symb. Log.*, 89(3):1313–1338, 2024.
13. Rodney G. Downey and Alexander G. Melnikov. Computably compact metric spaces. *Bull. Symb. Log.*, 29(2):170–263, 2023.
14. Mathieu Hoyrup, Takayuki Kihara, and Victor Selivanov. Degree spectra of homeomorphism types of compact Polish spaces. *The Journal of Symbolic Logic*. Published online 2023:1–32.
15. Martino Lupini, Alexander Melnikov, and Andre Nies. Computable topological abelian groups. *J. Algebra*, 615:278–327, 2023.
16. Alec Fox. Effective metric structure theory of C*-algebras. PhD thesis, University of California, Irvine, 2023.
17. Tobias Fritz, Tim Netzer, and Andreas Thom. Can you compute the operator norm? *Proceedings of the American Mathematical Society*, 142(12):4265–4276, 2014.
18. Alexander V. Kuznetsov. Algorithms as operations in algebraic systems. *Uspekhi Matematicheskikh Nauk*, 13(3):81, 1958.
19. Joachim Cuntz. Simple C*-algebra generated by isometries. *Communications in Mathematical Physics*, 57(2):173–185, 1977.
20. Kenneth R. Davidson. C*-algebras by Example. Fields Institute Monographs. American Mathematical Society, Providence, R.I., 1996.
21. Alexander Lubotzky and Yehuda Shalom. Finite representations in the unitary dual and Ramanujan groups. *Contemporary Mathematics*, 347:173–190, 2004.
22. Anatolii I. Mal'tsev. On isomorphic matrix representations of infinite groups. *Rec. Math.[Mat. Sbornik] NS*, 8(50):405–422, 1940.
23. MB Bekka. On the full C*-algebras of arithmetic groups and the congruence subgroup problem. *Forum Mathematicum*, 11:705–715, 1999.
24. Eberhard Kirchberg. On non-semisplit extensions, tensor products and exactness of group C*-algebras. *Inventiones mathematicae*, 112:449–489, 1993.
25. Zhengfeng Ji, Anand Natarajan, Thomas Vidick, John Wright, and Henry Yuen. MIP*=RE. *Communications of the ACM*, 64(11):131–138, 2021.
26. Terry A. Loring. *Lifting solutions to perturbing problems in C*-algebras*, volume 8 of *Fields Institute Monographs*. American Mathematical Society, Providence, RI, 1997.

27. Alec Fox, Isaac Goldbring, and Bradd Hart. Locally universal C*-algebras with computable presentations. *Journal of Functional Analysis*, 287(12):110652, 2024.
28. Isaac Goldbring and Thomas Sinclair. On Kirchberg's embedding problem. *Journal of Functional Analysis*, 269(1):155–198, 2015.
29. Bruce Blackadar. Operator algebras: Theory of C*-algebras and von Neumann algebras, volume 122. Springer Science & Business Media, 2006.
30. Francis J. Murray and John von Neumann. On rings of operators. IV. *Annals of Mathematics*, 44(4):716–808, 1943.
31. Isaac Goldbring. Oracle computability of conditional expectations onto subfactors. *New York Journal of Mathematics*, 27:1085–1095, 2021.
32. Ilijas Farah, Isaac Goldbring, Bradd Hart, and David Sherman. Existentially closed II_1 factors. *Fundamenta Mathematicae*, 233:173–196, 2017.
33. Vaughan FR Jones. Index for subfactors. *Inventiones mathematicae*, 72(1):1–25, 1983.
34. Alec Fox and Isaac Goldbring. Effective subfactor theory. *arXiv preprint* arXiv:2409.18831, 2024.
35. Marian B. Pour-El and J. Ian Richards. *Computability in Analysis and Physics*. Perspectives in Mathematical Logic. Springer, Berlin, 1989.
36. Alexander G. Melnikov and Keng Meng Ng. Computable structures and operations on the space of continuous functions. *Fund. Math.*, 233(2):101–141, 2016.
37. Nikolay Bazhenov, Matthew Harrison-Trainor, and Alexander Melnikov. Computable Stone spaces. *Ann. Pure Appl. Logic*, 174(9):Paper No. 103304, 25, 2023.

Chapter 8
Degrees of Theories

In this chapter, we consider the computability-theoretic complexities of the theories of various metric structures. In the first section, we consider the complexity of the diagram of a metric structure and establish results for first-order formulas as well as for infinitary formulas. We then study the complexity of the theory of a structure. As alluded to in Chap. 2, there are two perspectives on the theory of a metric structure, and these two perspectives yield slightly different effective results. Afterwards, we consider a particular kind of structure arising in model theory, the *finitely generic* structures, and establish upper bounds on the complexities of their theories. The final section studies the universal theory of the hyperfinite II_1 factor, whose undecidability is intimately connected with a recent result of quantum complexity theory known as $\mathrm{MIP}^* = \mathrm{RE}$.

8.1 Complexities of Diagrams

8.1.1 Finitary Results

In classical logic, every formula is equivalent to one in prenex normal form. Thus, in discussing the complexity of the theory of a structure $\mathcal{M}$, it makes sense to focus attention on the computability-theoretic complexity of the set of sentences true in $\mathcal{M}$ of a fixed quantifier-theoretic complexity.

For example, suppose that $\mathcal{M}$ is a computably presentable *classical* L-structure where L is a computable language, and let us consider the universal theory $\mathrm{Th}_\forall(\mathcal{M})$ of $\mathcal{M}$, that is, $\mathrm{Th}_\forall(\mathcal{M})$ is the set of all Π_1 L-sentences that are true in $\mathcal{M}$. It is not too hard to see that $\mathrm{Th}_\forall(\mathcal{M})$ is Π^0_1.[1] Indeed, $\sigma = \forall x \varphi(x)$ belongs to the universal theory

[1] Of course, one is really talking about the subset of the set of Gödel codes for universal L-sentences (which is a computable set) that code sentences true in $\mathcal{M}$.

J. N. Y. Franklin et al., *Effective Metric Structure Theory*, Springer Monographs in Mathematics, https://doi.org/10.1007/978-3-032-13322-9_8

of $\mathcal{M}$ if and only if, for all $a \in M$, one has $\mathcal{M} \models \varphi(a)$. If $\mathcal{M}^\#$ is a computable presentation of $\mathcal{M}$, then $\{a \in M : \mathcal{M} \models \varphi(a)\}$ is a computable set of $\mathcal{M}^\#$. Of course, in some instances, the universal theory of $\mathcal{M}$ (or even the entire theory of M) can be computable. For example, the theory of $(\mathbb{Q}, <)$ is computable. This can be demonstrated by an effective elimination of quantifiers argument. Thus, we are really only speaking of upper bounds at the moment.

This alignment between quantifier complexity and computability-theoretic complexity remains true at higher levels. Namely, for any n, the set of Π_n (respectively Σ_n) sentences true in $\mathcal{M}$ defines a Π^0_n (respectively Σ^0_n) set. The proof proceeds by induction and resembles the case of $n = 1$ treated in the previous paragraph.

In this section, we discuss analogous results in the setting of metric structures. A few preliminary remarks are in order as to what this entails. Fix a continuous signature L and an L-structure $\mathcal{M}$. Unlike the discrete case, some L-formulas may not be equivalent to one in prenex normal form. However, the set of L-formulas in prenex normal form is dense in the set of all L-formulas [1, Sect. 6]. Therefore, it still makes sense to restrict our attention to such formulas. However, given that interpretations of sentences in metric structures now take real values, one must specify exactly the sense in which one wants to "compute" these various fragments of the theory of $\mathcal{M}$. We present one such approach in this section. All results in this section are due to Camrud, Goldbring, and McNicholl [2].

Throughout this section, we assume that $\mathcal{M}$ is a computably presentable (metric) L-structure where L is a computable language. For simplicity, we restrict our attention to the case where all predicate symbols (including the metric) take values in $[0, 1]$ and our structure is 1-sorted. Just as in classical logic, one may define the collections of finitary $\forall_n$ and $\exists_n$ formulas, the only difference being that universal quantifiers are replaced with sup's, existential quantifiers are replaced with inf's, and the quantifier-free formulas are assumed to be restricted (see the remarks after Definition 2.9). In what follows, for a restricted sentence σ, we abuse notation and write $\sigma \in \Sigma_n$ to mean that σ is a Σ_n-sentence and likewise with Π_n.

For each $n \geq 1$, $X \in \{\Sigma, \Pi\}$, and $\bowtie \in \{<, \leq\}$, we set

$$\mathrm{Diag}_{n,X,\bowtie}(\mathcal{M}) = \{(\sigma, q) : \sigma \in X_n, q \in \mathbb{Q} \cap [0, 1], \sigma^M \bowtie q\}.$$

Each of the above sets is a kind of *diagram* for $\mathcal{M}$.

Theorem 8.1 *Let n be a positive integer.*

1. $\mathrm{Diag}_{n,\Pi,\leq}(\mathcal{M})$ *is* Π^0_n.
2. $\mathrm{Diag}_{n,\Pi,<}(\mathcal{M})$ *is* Σ^0_{n+1}.
3. $\mathrm{Diag}_{n,\Sigma,\leq}(\mathcal{M})$ *is* Π^0_{n+1}.
4. $\mathrm{Diag}_{n,\Sigma,<}(\mathcal{M})$ *is* Σ^0_n.

Moreover, these results are uniform in the sense that from n and an index for a computable presentation of $\mathcal{M}$, one can compute an index for any of the above diagrams.

The idea behind the proof is quite similar to the classical case, and we simply sketch the idea for $n = 1$. Fix $\sigma = \sup_x \varphi(x)$ with φ quantifier free and $q \in \mathbb{Q} \cap [0, 1]$. Given a rational point a from the presentation for $\mathcal{M}$ and $m \geq 1$, it is decidable whether or not $\varphi^M(a) < q + \frac{1}{m}$. Consequently, $(\sigma, q) \in \mathrm{Diag}_{1,\Pi,\leq}$ if and only if, for all a from the presentation and all $m \geq 1$, we have $\varphi^M(a) < q + \frac{1}{m}$; this is clearly a Π^0_1 set. It then follows that $\sigma^M < q$ if and only if there is $q' < q$ such that $\sigma^M \leq q'$; this is now seen to be a Σ^0_2 set. The proof for Σ_1 sentences is similar, and the general case is handled by induction.

Although Theorem 8.1 is stated in terms of upper bounds for sentences, analogous results can easily be obtained for lower bounds. For example, if σ is a Σ_n-sentence, then $1 \dot{-} \sigma$ is equivalent to a Π_n-sentence. Since $\sigma^M \geq r$ is equivalent to $(1 \dot{-} \sigma)^M \leq 1 \dot{-} r$, we see that checking whether or not $\sigma^M \geq r$ is a Π^0_n-condition.

It turns out that not only are the bounds in Theorem 8.1 optimal, but there is a single structure that witnesses the optimality of all of these bounds simultaneously.

Theorem 8.2 *There is a computable language L and a computably presentable L-structure $\mathcal{M}$ such that all of the above diagrams are complete for their respective levels of the arithmetic hierarchy.*

Proof Let L be the language that consists of the following:

1. A constant symbol $\underline{0}$.
2. A family of unary predicate symbols $(C_n)_{n \in \mathbb{N}}$.
3. A family of predicate symbols $(P_{N,n})_{N,n \in \mathbb{N}}$ where $P_{2N,n}$ and $P_{2N+1,n}$ are $(N + 1)$-ary.

Each of the predicate symbols has modulus of continuity $n \mapsto 1$. (Here, we define modulus of continuity as in Chap. 3).

We now define our structure $\mathcal{M}$. The domain of $\mathcal{M}$ is simply $\mathbb{N}$, and we set $\underline{0}^{\mathcal{M}} = 0$. In order to define the interpretations of the predicates, we first introduce some notation. For $N \in \mathbb{N}$ and $R \subseteq \mathbb{N}^{N+1}$, we set:

$$
\begin{aligned}
\boldsymbol{\exists} R &= \{n \in \mathbb{N} \ : \ \exists x_1 \forall x_2 \ldots Q x_N \ R(n, x_1, \ldots, x_N)\} \\
\boldsymbol{\forall} R &= \{n \in \mathbb{N} \ : \ \forall x_1 \exists x_2 \ldots Q x_N \ R(n, x_1, \ldots, x_N)\} \\
R^* &= \{(n, x_1, \ldots, x_N) \ \in \mathbb{N}^{N+1} \ : \ (\forall x_1' \leq x_1) \\
&\qquad (\exists x_2' \leq x_2) \ldots (Q x_N' \leq x_N) R(n, x_1', \ldots, x_N')\} \neg R = \mathbb{N}^{N+1} \setminus R
\end{aligned}
$$

Note that R is Turing equivalent to R^*.

We then fix a uniformly computable family $(R_N)_{N \in \mathbb{N}}$ of relations so that for each $N \in \mathbb{N}$, $R_{2N} \cup R_{2N+1} \subseteq \mathbb{N}^{N+2}$, $\boldsymbol{\forall} R_{2N}$ is Π^0_{N+1}-complete, and $\boldsymbol{\exists} R_{2N+1}$ is Σ^0_{N+1}-complete.

For $K, N \in \mathbb{N}$ and $f : \mathbb{N}^{N+1} \to \mathbb{R}$ a bounded function, set:

$$
\Gamma_K(f; x_1, \ldots, x_N) = \sum_{x_0=0}^{K} 2^{-(x_0+1)} f(x_0, \ldots, x_N)
$$

$$\Gamma(f; x_1, \ldots, x_N) = \sum_{x_0=0}^{\infty} 2^{-(x_0+1)} f(x_0, \ldots, x_N).$$

We define $\Gamma(f) : \mathbb{N}^N \to \mathbb{R}$ by setting $\Gamma(f)(x_1, \ldots, x_N) = \Gamma(f; x_1, \ldots, x_N)$. We note that $\Gamma(f)$ is computable if f is computable and, in this case, an index of $\Gamma(f)$ can be computed from an index of f and a bound on f.

We then define

$$f_N = \begin{cases} \Gamma(1 - \frac{1}{2}\chi_{R_N^*}) & N \text{ even} \\ \Gamma(\frac{1}{2}\chi_{(\neg R_N)^*}) & \text{otherwise.} \end{cases}$$

Also, for every $a \in \mathbb{N}$, set

$$C_n^{\mathcal{M}}(a) = \begin{cases} f_0(n/2) & n \text{ even} \\ f_1((n-1)/2) & \text{otherwise.} \end{cases}$$

Finally, set $P_{2N,n}^{\mathcal{M}}(a_0, \ldots, a_N) = f_{2N+2}(a_0, \ldots, a_N, n)$ and $P_{2N+1,n}^{\mathcal{M}}(a_0, \ldots, a_N) = f_{2N+3}(a_0, \ldots, a_N, n)$.

Having defined our structure $\mathcal{M}$, we now define, for each $N, n \in \mathbb{N}$, a Σ_n-sentence $\phi_{N,n}$ of L and a Π_n-sentence $\psi_{N,n}$ of L. To begin, we set:

$$\phi_{0,n} = C_{2n}(\underline{0})$$
$$\psi_{0,n} = C_{2n+1}(\underline{0}).$$

Then, for each positive integer N, we set:

$$\phi_{N,n} = \inf_{x_1} \ldots Q_{x_N} P_{2N,n}(x_1, \ldots, x_N)$$
$$\psi_{N,n} = \sup_{x_1} \ldots Q_{x_N} P_{2N+1,n}(x_1, \ldots, x_N).$$

The bulk of the proof is demonstrating the following equivalences for $n, N \in \mathbb{N}$ with $N > 0$. We refer the reader to [2] for the details.

$$\phi_{0,n}^{\mathcal{M}} \leq \frac{1}{2} \Leftrightarrow n \in \forall R_0$$
$$\psi_{0,n}^{\mathcal{M}} < \frac{1}{2} \Leftrightarrow n \in \exists R_0$$
$$\phi_{N,n}^{\mathcal{M}} \leq \frac{1}{2} \Leftrightarrow n \in \forall R_{2N}$$
$$\psi_{N,n}^{\mathcal{M}} < \frac{1}{2} \Leftrightarrow n \in \exists R_{2N+1}$$

□

We note that the language L constructed in the proof of Theorem 8.2 has infinitely many predicate symbols. We are thus led to the following.

Question 8.3 Is there a language L with finitely many predicate symbols and a computably presentable L-structure $\mathcal{M}$ that achieves all of the upper bounds in Theorem 8.1?

We will present a likely candidate below in Sect. 8.4.

8.1.2 Infinitary Results

We now consider a version of the results presented in the previous subsection in the context of infinitary formulas, that is, formulas obtained by allowing countable infima and suprema (the continuous analogs of countable disjunction and conjunction). Such formulas are said to belong to the collection of $L_{\omega_1\omega}$*-formulas*. The $L_{\omega_1\omega}$ formulas in *normal form* are defined by transfinite recursion on $\alpha < \omega_1$. As in the case of finitary formulas, Σ_0 and Π_0 both denote the set of quantifier-free restricted (finitary) formulas. For $\alpha > 0$, one defines the set of Σ_α formulas to be the formulas of the form $\bigwedge_n \psi_n$ where each ψ_n is of the form $\inf_{\overline{z_n}} \varphi_n$, with φ_n a Π_{β_n} formula for some $\beta_n < \alpha$. Here, following usual lattice-theoretic notation, $\bigwedge$ denotes the operation of taking the infimum. Note that taking the infimum acts like taking the disjunction in that $\bigwedge_n \psi_n = 0$ if and only if some $\psi_n = 0$ (approximately). Thus, the lattice-theoretic notation is the opposite of the usual logical notation $\bigvee$ for an infinite disjunction. One defines the set of Π_α formulas analogously. Namely, $\bigvee$ denotes the operation of taking the supremum, and a Π_α formula is one of the form $\bigvee_n \psi_n$ where each ψ_n is of the form $\sup_{\overline{z_n}} \varphi_n$, with φ_n a Σ_{β_n} formula for some $\beta_n < \alpha$.

Suppose now that L is a computable language. As in the classical case, specifying which $L_{\omega_1\omega}$ formulas are computable takes some care, the point being that one should only take "effective" infima and suprema. To describe such formulas precisely, one needs to use the notions of computable ordinals and Kleene's $\mathcal{O}$ (see Appendix B).

We define the *computable formulas of* $L_{\omega_1\omega}$ and their respective *codes* by effective transfinite recursion. Our treatment follows the development of the classical case in [3]. We begin by fixing an effective enumeration of the quantifier-free restricted L-formulas as well as effective codings of the following.

1. All pairs of the form $(j, \overline{z})$ where $j \in \mathbb{N}$ and $\overline{z}$ is a tuple of variables.
2. All quadruples of the form $(X, a, \overline{x}, e)$ where $X \in \{\Sigma, \Pi\}$, $a, e \in \mathbb{N}$ and $\overline{x}$ is a tuple of variables.

When ξ is a tuple of either of the above types, we let $\overline{\xi}$ denote the code of ξ.

For every $X \in \{\Sigma, \Pi\}$ and $a \in \mathcal{O}$, we first define the index set S_a^X in such a way that if a is a notation for α, then every formula with indices in S_a^X will belong to the set of X_α formulas.

We begin by setting $S_1^\Sigma = S_1^\Pi$ to be the set of codes of all quantifier-free, restricted (finitary) formulas of L. (Recall that 1 denotes the ordinal 0 in $\mathcal{O}$.) For every $a \in$

$\mathcal{O} \setminus \{1\}$ and $X \in \{\Sigma, \Pi\}$, let S_a^X be the set of codes of all quadruples of the form $(X, a, \overline{x}, e)$ where $\overline{x}$ is a finite tuple of variable symbols and $e \in \mathbb{N}$.

Now for every $a \in \mathcal{O}$, $X \in \{\Sigma, \Pi\}$, and tuple of variable symbols $\overline{x}$, we define $P(X, a, \overline{x})$ to be the set of all codes of pairs $(j, \overline{z})$ where j codes a quadruple $(X, b, \overline{y}, e')$ with $b <_{\mathcal{O}} a$ and $\overline{z}$ is a finite sequence of variable symbols from $\overline{y}$ not contained in $\overline{x}$.

For each $i \in S_a^\Sigma \cup S_a^\Pi$, we define an infinitary formula ϕ_i as follows. Recall that W_e denotes the domain of the eth partial computable function.

1. If $a = 1$, then ϕ_i is the quantifier-free finitary formula indexed by i.
2. Suppose $a > 1$ and $i = \overline{(X, a, \overline{x}, e)}$.

   a. If $X = \Sigma$, then set
   $$\phi_i = \bigwedge_{\overline{(j,\overline{z})} \in W_e \cap P(\Pi, a, \overline{x})} \inf_{\overline{z}} \phi_j .$$

   b. If $X = \Pi$, then set
   $$\phi_i = \bigvee_{\overline{(j,\overline{z})} \in W_e \cap P(\Sigma, a, \overline{x})} \sup_{\overline{z}} \phi_j .$$

For every computable ordinal α, we let Σ_α^c denote the set of all formulas ϕ_i where $i \in \bigcup_{a \in \langle \alpha \rangle} S_a^\Sigma$. Similarly, Π_α^c denotes the set of all formulas ϕ_i where $i \in \bigcup_{a \in \langle \alpha \rangle} S_a^\Pi$. If $\psi = \phi_i$, then we say that i is a code of ψ. By a computable infinitary formula, we mean an element of $\Sigma_\alpha^c \cup \Pi_\alpha^c$ for some computable ordinal α.

It is fairly routine to verify that all logical operations can be performed effectively via this coding system. For example, from an i that codes an infinitary formula ϕ, it is possible to compute a code of $\sup_x \phi$.

When formulating our diagram complexity results for infinitary logic, we need to refrain from using the terminology of diagrams. The reason for this is that, because of the coding of the computable infinitary formulas, these diagrams compute $\mathcal{O}$, which is a Π_1^1-complete set. In order to obtain more meaningful results, we focus on the complexity of the right Dedekind cuts of reals of the form ϕ^M where ϕ is a computable infinitary formula. Towards that end, for $x \in \mathbb{R}$ and $\bowtie \in \{>, \geq\}$, we set $D^{\bowtie}(x) = \{q \in \mathbb{Q} : q \bowtie x\}$.

As in the case of finitary sentences, we first establish upper bounds on complexity.

Theorem 8.4 *Suppose that $\mathcal{M}$ is a computably presentable structure in a computable language L and that σ is a computable infinitary L-sentence.*

1. *If $\sigma \in \Pi_\alpha^c$, then $D^>(\sigma^\mathcal{M})$ is $\Sigma_{\alpha+1}^0$ and $D^\geq(\sigma^\mathcal{M})$ is Π_α^0.*
2. *If $\sigma \in \Sigma_\alpha^c$, then $D^>(\sigma^\mathcal{M})$ is Σ_α^0 and $D^\geq(\sigma^\mathcal{M})$ is $\Pi_{\alpha+1}^0$.*

Moreover, an index for any of these sets can be computed uniformly from a code for σ.

Proof The proof of the theorem proceeds by induction on α. We treat one case as an example of how the argument works in general. Suppose that ϕ is a Π^c_α sentence. Thus, ϕ has the form $\bigvee_{i\in I} \sup_{\overline{x_i}} \phi_i$ where I is c.e. and ϕ_i is $\Sigma^c_{\beta_i}$ for some $\beta_i < \alpha$ uniformly in i. Let $\mathbb{Q}(M^\#)$ denote the set of rational points of $M^\#$. Then, for $q \in \mathbb{Q}$, we have

$$q \in D^{\geq}(\phi^M) \Leftrightarrow (\forall k \in \mathbb{N})(\forall i \in I)(\forall \overline{r} \in \mathbb{Q}(M^\#))\ q + 2^{-k} \in D^{>}(\phi_i^M(\overline{r})).$$

By the inductive assumption, $\emptyset^{(\alpha)}$ computes $D^{>}(\phi_i^M(\overline{r}))$ uniformly in i, whence $D^{\geq}(\phi^M)$ is co-c.e. in $\emptyset^{(\alpha)}$. Thus, $D^{\geq}(\phi^M)$ belongs to Π^0_α as desired. □

As in the finitary case, the upper bounds from the previous theorem are sharp. This can be demonstrated by applying the following result of Camrud and McNicholl [4].

Theorem 8.5 *Let L be the language of metric spaces. Suppose* $0 < \alpha < \omega_1^{\mathrm{ck}}$ *and let* $r \in [0, 1]$.

1. *If the right Dedekind cut of r is* Σ^0_α, *then there is a* Σ^c_α *L-sentence* Φ *so that* $\Phi^{\mathcal{M}} = r$ *for every L-structure* $\mathcal{M}$.
2. *If the left Dedekind cut of r is* Σ^0_α, *then there is a* Π^c_α *L-sentence* Φ *so that* $\Phi^{\mathcal{M}} = r$ *for every L-structure* $\mathcal{M}$.

Theorem 8.5 is proved by first observing that for every dyadic rational $r \in [0, 1]$, there is an L-sentence σ so that $\sigma^{\mathcal{M}} = r$ for every L-structure $\mathcal{M}$. The required sentences are then constructed by transfinite recursion.

8.2 Complexities of Theories

As mentioned in Chap. 2, there are two perspectives on theories in continuous logic. When it comes to describing the complexity of a theory, these two perspectives differ slightly. Throughout this section, fix a continuous language L and an L-structure $\mathcal{M}$.

On the one hand, we may view the complete theory of $\mathcal{M}$ as the *function* $\mathrm{Th}(\mathcal{M})$, whose domain is the set of L-sentences and which is defined by setting $\mathrm{Th}(\mathcal{M})(\sigma) = \sigma^M$. Given a Gödel coding of restricted L-sentences, we can view $\mathrm{Th}(\mathcal{M})$ as a function $\mathrm{Th}(M) : \mathbb{N} \to \mathbb{R}$. Given a degree **d**, we say $\mathrm{Th}(\mathcal{M})$ is **d**-computable if the function $\mathrm{Th}(\mathcal{M}) : \mathbb{N} \to \mathbb{R}$ is a **d**-computable function; that is, if there is a **d**-algorithm such that, given a code for a restricted sentence σ and $k \in \mathbb{N}$, computes $q \in \mathbb{Q}$ so that $|q - \sigma^{\mathcal{M}}| < 2^{-k}$. Given $n \geq 1$ and $X \in \{\Sigma, \Pi\}$, we can consider the restriction $\mathrm{Th}_{X_n}(\mathcal{M})$ of $\mathrm{Th}(\mathcal{M})$ to the (computable) set of X_n-sentences. Note also that, using the "weak negation" $\sigma \mapsto 1 \dot{-} \sigma$, $\mathrm{Th}_{\Sigma_n}(\mathcal{M})$ is **d**-computable if and only if $\mathrm{Th}_{\Pi_n}(\mathcal{M})$ is **d**-computable. Thus, to simplify the discussion that follows, we simply worry about the complexity of $\mathrm{Th}_{\Pi_n}(\mathcal{M})$.

The discussion in the preceding section yields this theorem.

Theorem 8.6 *Suppose that $\mathcal{M}$ has a* **d***-computable presentation. Then* $\mathrm{Th}_{\Pi_n}(\mathcal{M})$ *is* $\mathbf{d}^{(\mathbf{n})}$*-computable and* $\mathrm{Th}(\mathcal{M})$ *is* $\mathbf{d}^{(\omega)}$*-computable.*

We give some sample applications of this theorem from operator algebras.

Example 8.7 Suppose that G is a finitely presented group. As mentioned in Corollary 7.19, the standard presentation of $C^*(G)$ is $\mathbf{0}'$-computable. Consequently, $\mathrm{Th}_{\forall_n}(C^*(G))$ is $\mathbf{0}^{(n+1)}$-computable and $\mathrm{Th}(C^*(G))$ is $\mathbf{0}^{(\omega)}$-computable.

Example 8.8 Suppose that G is a finitely generated group, and let **d** be a Turing degree that computes the word problem of G. By relativizing Lemma 7.20, the standard presentation of $C_r^*(G)$ is $\mathbf{d}'$-computable. In this case, $\mathrm{Th}_{\forall_n}(C^*(G))$ is $\mathbf{d}^{(\mathbf{n}+1)}$-computable and $\mathrm{Th}(C_r^*(G))$ is $\mathbf{d}^{(\omega)}$-computable. In particular, if G has a solvable (or even arithmetic) word problem, then $\mathrm{Th}(C_r^*(G))$ is $\mathbf{0}^{(\omega)}$-computable.

Example 8.9 Suppose that G is a finitely generated group with a word problem solvable from the Turing degree **d**. By relativizing Example 7.46, the standard presentation of $L(G)$ is **d**-computable. Consequently, $\mathrm{Th}_{\forall_n}(L(G))$ is $\mathbf{d}^{(\mathbf{n})}$-computable and $\mathrm{Th}(L(G))$ is $\mathbf{d}^{(\omega)}$-computable. In particular, if G has a solvable (even arithmetic) word problem, then $\mathrm{Th}(L(G))$ is $\mathbf{0}^{(\omega)}$-computable.

Example 8.10 For any of the C*-algebras A shown to have a computable presentation, such as $\mathcal{O}_2$ or A_θ for θ a computable irrational real, we have that $\mathrm{Th}_{\forall_n}(A)$ is $\mathbf{0}^{(n)}$-computable and $\mathrm{Th}(A)$ is $\mathbf{0}^{(\omega)}$-computable.

On the other hand, we can view a (not necessarily complete) theory T as simply a set of codes for restricted L-sentences, whence it makes sense, for example, to say that T is Σ_n^0. Recall that a theory T (in this sense) is complete if and only if, for every restricted L-sentence σ, there is a unique $r \in \mathbb{R}$ such that $|\sigma - r| \in T$. Note, then, that T is complete if and only if there is a structure $\mathcal{M}$ such that $T = \mathrm{Th}(\mathcal{M})^{-1}(\{0\})$. As a result, we have the following.

Theorem 8.11 *For any degree* **d***, we have that* $\mathrm{Th}(M)$ *is* **d***-computable if and only if the associated theory* $\mathrm{Th}(\mathcal{M})^{-1}[\{0\}]$ *is* Π_1^0 *in* **d**.

As mentioned in Sect. 2.3, if T is a c.e. L-theory, then the set of restricted L-sentences σ such that $T \vdash \sigma$ is also c.e. The following is a straightforward consequence of this fact and Theorem 2.18.

Theorem 8.12 *If T is a c.e. L-theory, then for any restricted L-sentence σ, we have that* $\sup\{\sigma^{\mathcal{M}} : \mathcal{M} \models T\}$ *is a right c.e. real uniformly in a code for σ.*

Another straightforward consequence of the above observations is the following.

Theorem 8.13 *Suppose that $\mathcal{M}$ is an L-structure for which there is a c.e. L-theory T equivalent to* $\mathrm{Th}(\mathcal{M})^{-1}[\{0\}]$*. Then* $\mathrm{Th}(\mathcal{M})$ *is computable.*

Proof Suppose a restricted L-sentence σ and $k \in \mathbb{N}$ are given. Since T is c.e., the set of its consequences is c.e. We then search the consequences of T for one of the form $|\sigma - s| \dot{-} \epsilon$ where $s, \epsilon \in \mathbb{Q}$ and $\epsilon < 2^{-k}$. Once such a consequence is found, we output s.

The correctness of this algorithm follows from the soundness of the proof system. Namely if $T \vdash |\sigma - s| \dot{-} \epsilon$, then since $\mathcal{M} \models T$, $(|\sigma - s| \dot{-} \epsilon)^{\mathcal{M}} = 0$, which is just another way to say that $|\sigma^{\mathcal{M}} - s| \leq \epsilon$. The only thing left to show is that this algorithm halts on all σ, k. To this end, we only have to show that T has a consequence of the form $|\sigma - s| \dot{-} \epsilon$ where $\epsilon < 2^{-k}$. Set $\epsilon = 2^{-(k+1)}$, and let s be any positive rational number such that $|\sigma^{\mathcal{M}} - s| < \epsilon/2$. Then, since T is equivalent to $\mathrm{Th}(\mathcal{M})^{-1}(0)$, it follows from Theorem 2.18 that

$$\inf\left\{\delta \in \mathbb{Q}^{>0} \ : \ T \vdash \left(|\sigma - s| \dot{-} \frac{\epsilon}{2}\right) \dot{-} \delta\right\} = \sup\left\{\left(|\sigma - s| \dot{-} \frac{\epsilon}{2}\right)^{\mathcal{N}} \ : \ \mathcal{N} \models T\right\} = 0.$$

Hence, $T \vdash |\sigma - s| \dot{-} \epsilon$. □

8.3 Finite Forcing Companions

In this section, we consider a particular construction of first-order theories for which the complexity of the theory is rather predictable. The material in this section follows the paper [5] of Goldbring and Hart, which itself is partially modeled on the exposition of the classical version of the material presented in [6]. In addition, many of the facts listed here are proven in the paper [7] of Goldbring.

Fix a continuous language L and an L-theory T. We will construct a new theory, called the *finite forcing companion of* T, using a certain two-player game. To describe this game, we first set $L(C)$ to be the language obtained by adding a countable set $C = \{c_1, c_2, \ldots\}$ of new constant symbols to L. A *precondition* is a finite set p of expressions of the form $\varphi < r$ where φ is a restricted quantifier-free $L(C)$-sentence and r is a positive rational number. A precondition p is a *condition* (for T) if $T \cup p$ is satisfiable.

We consider a two-person game involving the players $\forall$ (for $\forall$belard) and $\exists$ (for $\exists$loise). Players $\forall$ and $\exists$ take turns playing conditions, where each player's move is required to extend the previous player's move. Thus, $\forall$ starts by playing the condition p_0, then $\exists$ plays a condition $p_1 \supseteq p_0$, then $\forall$ follows that play with some condition $p_2 \supseteq p_1$, etc. After ω many rounds, the two players have together played a chain $p_0 \subseteq p_1 \subseteq p_2 \subseteq \cdots$ of conditions whose union we will denote by $\bar{p}$.

We call the above play *definitive* if, for every atomic $L(C)$-sentence φ, there is a unique $r \in [0, 1]$ such that $T \cup \bar{p} \models \varphi = r$. In this case, $\bar{p}$ describes an $L(C)$-prestructure $A_0^+(\bar{p})$ whose completion will be denoted by $A^+(\bar{p})$ and will be referred to as the *compiled structure*. The reduct of $A^+(\bar{p})$ to L will be denoted by $A(\bar{p})$. If $\bar{p}$ is clear from context, we will denote $A^+(\bar{p})$ and $A(\bar{p})$ simply by A^+ and A, respectively.

Note that, regardless of player $\forall$'s moves, player $\exists$ can always ensure that the play of the game is definitive.

Definition 8.14 Let P be a property of $L(C)$-structures.

1. The game $G(P)$ is the game whose moves are as above and such that the player $\exists$ wins $G(P)$ if and only if $\bar{p}$ is definitive and $A^+(\bar{p})$ has property P.
2. We say that P is *enforceable* if the player $\exists$ has a winning strategy for $G(P)$.
3. We say that the condition p *forces* P, denoted $p \Vdash P$, if, for any position $(p_0, \ldots, p_k)$ of the game $G(P)$, if $p \subseteq p_k$, then the position is winning for $\exists$ in $G(P)$. We write $\Vdash P$ when $\emptyset \Vdash P$.

Fact 8.15 (Conjunction Lemma) If p forces P_i for each $i < \omega$, then p forces the conjunction $\bigwedge_{i<\omega} P_i$.

The set of *universal consequences of* T is the theory $T_\forall$ consisting of all nonnegative universal sentences σ such that $T \models \sigma$.

Fact 8.16 It is enforceable that the compiled structure be a model of $T_\forall$.

For a condition p and an $L(C)$-sentence φ, we set

$$F_p(\varphi) = \inf\{r \ : \ p \Vdash \varphi < r\}.$$

Fact 8.17 For any condition p and $L(C)$-sentence φ, we have $F_p(\varphi) \leq r$ if and only if $p \Vdash \varphi \leq r$.

We will require the following fact about the function F_p, whose proof can be found in [7, Fact 2.21 and Theorem 2.22].

Fact 8.18 For any condition p and $L(C)$-formula $\varphi(x)$, we have

$$F_p\left(\inf_x \varphi(x)\right) = \sup_{q \supseteq p} \inf_{q' \supseteq q} \inf_{c \in C} F_{q'}(\varphi(c)).$$

Suppose now that L is a computable language and that T is an L-theory that is c.e. in $\mathbf{d}$ for some Turing degree $\mathbf{d}$. We let Pre_T denote the set of all codes of preconditions. More formally, a tuple $(e_1, \ldots, e_n)$ belongs to Pre_T if for each e_i there exist k, r such that k codes a quantifier-free restricted $L(C)$-sentence, $r \in \mathbb{Q}^{>0}$, and e_i codes (k, r). We let Cond_T denote those $e \in \mathrm{Pre}_T$ that code a condition for T. If $e \in \mathrm{Pre}_T$, we let p_e denote the precondition it codes.

Lemma 8.19

1. Pre_T *is* $\mathbf{d}'$*-computable.*
2. *The set* $\{(e_1, e_2) \in \mathrm{Pre}_T \ : \ p_{e_1} \subseteq p_{e_2}\}$ *is* $\mathbf{d}$*-computable.*
3. Cond_T *is co-c.e. in* $\mathbf{d}$.

The first two statements are clear, while the last statement follows from the fact that $\{\varphi_{k_i} < r_i \ : \ i = 1, \ldots, n\}$ is a condition for T if and only if $T \nvDash \min_{i=1,\ldots,n} r_i \dot{-} \varphi_{k_i}$.

Given a precondition $p = \{\varphi_i < r_i \ : \ i = 1, \ldots, n\}$, let

$$p^{\$} = \bigcup \left\{ \max_{1 \leq i \leq n} (\varphi_i \dot{-} s_i) \ : \ s_1, \ldots, s_n \in \mathbb{Q}, \ s_i < r_i \right\}.$$

Note that $p^{\$}$ is an $L(C)$-theory such that $A \models T \cup p$ if and only if there is $\theta \in p^{\$}$ such that $A \models T \cup \theta$.

We use Quine corners $\ulcorner \cdot \urcorner$ for Gödel numbers of formulas. The following lemma is clear.

Lemma 8.20 *There is a computable set $E \subseteq \mathbb{N}^2$ such that $(e, c) \in E$ if and only if $e \in \mathrm{Pre}_T$, c is the code of an $L(C)$-sentence, and $\varphi_c \in p^{\$}$.*

Proposition 8.21 *Suppose that p is a condition for T and $\psi(x)$ is a quantifier-free $L(C)$-formula. Then the following statements are equivalent.*

1. *$p \Vdash \sup_x \psi(x) \leq r$.*
2. *For all $\theta \in p^{\$}$, $T \cup \{\theta\} \vdash \sup_x \psi(x) \dot{-} r$.*

Proof First suppose that $\theta \in p^{\$}$ is such that $T \cup \{\theta\} \nvDash (\sup_x \psi(x)) \dot{-} r$. By Theorem 2.18, there is an $L(C)$-structure A such that $A \models T \cup \{\theta\}$ and $(\sup_x \psi(x))^A > r$. Take constants c from C such that $\psi(c)^A > r$. Set $q = p \cup \{1 \dot{-} \psi(c) < 1 - r\}$ and note that it is a condition for which $q \Vdash \psi(c) > r$. Consequently, $p \nVdash \sup_x \psi(x) \leq r$.

Conversely, suppose that $T \cup \{\theta\} \vdash (\sup_x \psi(x)) \dot{-} r$ for all $\theta \in p^{\$}$. Suppose that player $\forall$ plays p. Then player $\exists$ can play so that the compiled structure A is such that $A \models T_\forall \cup p$. Take $B \models T$ such that $A \subseteq B$, and take $\theta \in p^{\$}$ such that $B \models \theta$. By assumption and Theorem 2.18 again, $(\sup_x \psi(x))^B \leq r$, whence $(\sup_x \psi(x))^A \leq r$. It follows that $p \Vdash (\sup_x \psi(x)) \leq r$ as desired. □

The following theorem characterizes the complexity of the forcing relation.

Theorem 8.22 *For each $n \geq 0$, the set of triples (p, k, r) such that $e \in \mathrm{Cond}_T$ and k is the code for an $\forall_{2n+1}$-$L(C)$ sentence φ for which $F_{p_e}(\varphi) \leq r$ is Π^0_{2n+2} in* **d**.

Proof One proves the theorem by induction on n. The case $n = 0$ is immediate from the previous proposition. Suppose that $n \geq 1$ and $\varphi = \sup_x \inf_y \psi(x, y)$ where ψ is a $\forall_{2n-1}$-$L(C)$ sentence. Suppose further that p is a condition for T. Then the following are equivalent.

- $F_p(\varphi) \leq r$.
- $p \Vdash \varphi \leq r$.
- $(\forall c \in C) \ p \Vdash \inf_y \psi(c, y) \leq r$.
- $(\forall c \in C) \ F_p(\inf_y \psi(c, y)) \leq r$.
- $(\forall c \in C)(\forall q \supseteq p)(\forall m \in \mathbb{N})(\exists q' \supseteq q)(\exists d \in C) F_{q'}(\psi(c, d)) \leq r + \frac{1}{m}$.

By induction, the "matrix" in the last bullet is Π^0_{2n} in the parameters. Since the Gödel code for ψ is computable from the Gödel code for φ, we see that the entire condition is Π^0_{2n+2} in $\mathbf{d}$ as desired. □

The theory T has the *joint embedding property* (JEP) if any two models of T can be embedded into a common model of T.

Fact 8.23 If T has the JEP, then for any L-sentence σ, there is a unique real number r_σ such that $\Vdash \sigma = r_\sigma$.

Definition 8.24 If T has the JEP, then the *finite forcing companion of* T is the complete theory T^f such that, for all restricted L-sentences σ and all $r \in \mathbb{Q}^{>0}$, we have $\sigma \leq r \in T^f$ if and only if $r_\sigma \leq r$ (and similarly for $\sigma \geq r$).

The following is the main result of this section.

Theorem 8.25 *Suppose that T has the JEP and is c.e. in $\mathbf{d}$. Then T^f is $\mathbf{d}^{(\omega)}$-computable.*

To prove the theorem, fix a code k for an $\forall_{2n+1}$-$L(C)$ sentence φ and $\epsilon > 0$. Given positive $r \in \mathbb{Q}$, using $\mathbf{d}^{(2n+2)}$, one can determine whether or not $F_\emptyset(\varphi) \leq r$; that is, whether or not $\Vdash \varphi \leq r$. If the answer is "yes," then one concludes that $\varphi^{T^f} \leq r$. If the answer is "no," then, since we know that $\Vdash \varphi = \varphi^{T^f}$, we have that $\varphi^{T^f} > r$. Thus, one runs this algorithm over all possible rational $r > 0$ and waits until one sees that φ^{T^f} belongs to an interval of radius ϵ.

A theory T' is *model complete* if any embedding between models of T' is elementary. We call a theory T' a *model companion* for T if $T_\forall = T'_\forall$ and T' is model complete. A model companion for T, should it exist, is unique up to equivalence. In case T is axiomatized by a set of $\forall_2$-sentences, then T' is the model companion of T if and only if the models of T' are precisely the existentially closed models of T (see Sect. 7.2.11 for the definition of existentially closed model).

Fact 8.26 If T has a model companion, then the model companion of T coincides with the forcing companion T^f.

Corollary 8.27 *Suppose that T is c.e. in $\mathbf{d}$ and has a model companion T'. Then T' is $\mathbf{d}^{(\omega)}$-computable.*

The forcing companion is an interesting theory to consider, even in the absence of a model companion. We give some examples of the finite forcing technique from operator algebras.

Example 8.28 Let T be the theory of tracial von Neumann algebras, which does not have a model companion [8]. T is c.e. and has the JEP. It follows that T^f is $\mathbf{0}^{(\omega)}$-computable. It is natural to ask: What is T^f? Is there a concrete model of T^f? At the moment, we do not know of a concrete model of this theory. Due to the negative solution to the CEP (see Sect. 8.4 below) and [7, Sect. 5], it follows that $T^f \neq \mathrm{Th}(\mathcal{R})$.

Example 8.29 Let T be the theory of C*-algebras, which does not have a model companion [9]. T is c.e. and has the JEP. It follows that T^f is $\mathbf{0}^{(\omega)}$-computable. Once again, we currently do not know a concrete model of T^f. As shown in [7, Sect. 5], the Kirchberg Embedding Problem discussed in Sect. 7.2.9 above has a positive solution if and only if $T^f = \mathrm{Th}(\mathcal{O}_2)$.

Example 8.30 Let T be the theory of commutative C*-algebras. T is c.e. and has the JEP. In this case, T has a model companion, namely $\mathrm{Th}(C(2^{\mathbb{N}}))$ [10],[2] and thus we conclude $\mathrm{Th}(C(2^{\mathbb{N}}))$ is $\mathbf{0}^{(\omega)}$-computable. This conclusion could also be reached using Theorem 8.6 and the fact that, by Theorem 7.11, $C(2^{\mathbb{N}})$ is computably presentable.

Since $\mathrm{Th}(C(2^{\mathbb{N}}))$ is arguably the nicest theory of an infinite-dimensional operator algebra, it is not outlandish to propose the following.

Question 8.31 Is $\mathrm{Th}(C(2^{\mathbb{N}}))$ computable?

Example 8.32 Let T be the theory of projectionless[3] commutative C*-algebras. In other words, T axiomatizes the algebras $C(X)$ for X a totally disconnected compact space. Then T is c.e. (see [11, Remark 1.2]) and has the JEP. In [7], using the main result of [11], it was shown that $T^f = \mathrm{Th}(C^*(\mathbb{P}))$, where $\mathbb{P}$ is the so-called *pseudoarc* [12]. It follows that $\mathrm{Th}(C(\mathbb{P}))$ is $\mathbf{0}^{(\omega)}$-computable. Any standard construction of the pseudoarc demonstrates it is computably compact (see, for example, [13]). Thus, this result also follows from Theorem 7.11.

The following examples are not strictly about operator algebras, but are in the same spirit.

Example 8.33 Suppose that T is the theory of Banach spaces. Then T is c.e. and has the JEP. In this case, T has a model companion, namely $\mathrm{Th}(\mathbb{G})$, where $\mathbb{G}$ is the so-called *Gurarij Banach space* [14]. Consequently, $\mathrm{Th}(\mathbb{G})$ is $\mathbf{0}^{(\omega)}$-computable.

There are noncommutative versions of Banach spaces known as *operator spaces* and *operator systems*. An operator space is simply a Banach subspace of a C*-algebra, while an operator system is an operator space closed under $*$ and containing the identity. Suppose that T is the theory of operator spaces (respectively operator systems). Then T is c.e. [15, Appendix B] and has the JEP. In [7, Sect. 5], using the main results of [16], it was shown that $T^f = \mathrm{Th}(\mathbb{NG})$ (respectively $T^f = \mathrm{Th}(\mathbb{GS})$), where $\mathbb{NG}$ is the so-called *noncommutative Gurarij space* [17] (respectively $\mathbb{GS}$ is the *Gurarij operator system* [16]). Consequently, we have that $\mathrm{Th}(\mathbb{NG})$ and $\mathrm{Th}(\mathbb{GS})$ are $\mathbf{0}^{(\omega)}$-computable.

Question 8.34 Does the Gurarij Banach space $\mathbb{G}$, noncommutative Gurarij space $\mathbb{NG}$, or Gurarij operator system $\mathbb{GS}$ have a computable presentation?

---

[2] As in the previous chapter, in the remainder of this chapter, $C(X)$ denotes $C(X; \mathbb{C})$ viewed as a C*-algebra.

[3] A unital C*-algebra is *projectionless* if the only projections it contains are 0 and 1.

8.4 The Universal Theory of the Hyperfinite II_1 Factor

Since the hyperfinite II_1 factor $\mathcal{R}$ is computably presentable, it follows from Theorem 8.1 that the halting problem computes its universal theory. However, as was shown by Goldbring and Hart, this bound is sharp [18].

Theorem 8.35 *The universal theory of $\mathcal{R}$ computes the halting problem, that is, any Turing degree that computes the universal theory of $\mathcal{R}$ must also compute the halting problem.*

We note that this is the only known natural example of a continuous metric structure that achieves the highest possible complexity at a given quantifier level. Before discussing the proof of Theorem 8.35, we first examine some of its many interesting consequences. The first of these, on the face of it, has nothing to do with computability.

Corollary 8.36 *The Connes Embedding Problem (CEP) has a negative solution. That is, there are II_1 factors that do not embed into an ultrapower of $\mathcal{R}$.*

To prove the previous corollary, fix a universal (restricted) sentence σ and note that, since the axioms for II_1 factors are c.e., we have that $\{q \in \mathbb{Q}^{>0} : T_{II_1} \vdash \sigma \dot{-} q\}$ is c.e. On the other hand, if the CEP were true, then $\sup\{\sigma^M : M \models T_{II_1}\} = \sigma^{\mathcal{R}}$. By Theorem 2.18, these two quantities coincide. Therefore, $\sigma^{\mathcal{R}}$ is a right-c.e. real uniformly in σ, and this contradicts Theorem 8.35. That the CEP would imply that $\mathrm{Th}_\forall(\mathcal{R})$ is computable was first observed by Goldbring and Hart in [19].

The proof of Theorem 8.35 above uses a result from quantum complexity theory known as MIP* = RE, which we will describe momentarily. It was observed in [20] that MIP* = RE implies a negative resolution of the CEP, using very nontrivial intermediate results. Indeed, MIP* = RE implies a negative resolution to a problem in quantum information theory known as Tsirelson's problem. By work of Fritz [21], Junge et al. [22], and Ozawa [23], Tsirelson's problem is known to be equivalent to a problem in C*-algebra theory known as Kirchberg's QWEP Problem, which in turn was shown to be equivalent to the CEP in a fundamental paper of Kirchberg [24]. The upshot of the above discussion is that one can bypass these highly nontrivial intermediate equivalences and proceed directly from MIP* = RE to the failure of the CEP.

Besides being a more direct route to the resolution of the CEP, the above line of reasoning lends itself to a "Gödelian refutation" of the CEP. Indeed, the proof of Corollary 8.36 immediately yields the following.

Corollary 8.37 *There is no c.e. extension of the theory of II_1 factors whose models all embed into an ultrapower of $\mathcal{R}$.*

The previous corollary allows one to prove some purely operator algebraic results such as the following, which is not obviously derivable from the "standard" refutation described above.

Corollary 8.38 *There is a sequence* $M_1, M_2, \ldots$ *of separable* II_1 *factors, none of which embed into an ultrapower of* $\mathcal{R}$*, and such that, for all* $i < j$*,* M_i *does not embed into an ultrapower of* M_j*.*

Corollary 8.39 *The collection of counterexamples to the CEP is not closed under ultraproducts.*

The proofs of the previous two results can be found in [18].

In [25], Theorem 8.35 was extended to "nontracial versions" of $\mathcal{R}$. Indeed, factors that are not tracial are either of type II_∞ (meaning they are the tensor product of a type II_1 factor with $\mathcal{B}(\mathcal{H})$), or else they are of type III, which are classified by a real parameter $\lambda \in [0, 1]$. For each $\lambda \in (0, 1]$, there is a unique hyperfinite type III_λ factor. For $\lambda \in (0, 1)$, this factor is denoted $\mathcal{R}_\lambda$, while for $\lambda = 1$, this factor is denoted $\mathcal{R}_\infty$. There is an appropriate computable language for studying type III factors equipped with a faithful, normal state. The following result was proven by Arulseelan, Goldbring, and Hart [25].

Corollary 8.40 *The universal theory of* $\mathcal{R}_\infty$ *computes the halting problem.*

It is a fact that, for any two faithful normal states φ and ψ on a type III_1 factor M, one has that (M, φ) and (M, ψ) are elementarily equivalent. For this reason, we may abuse terminology and omit mention of the state in the previous result. The key step in deriving the previous corollary from Corollary 8.37 above is that a tracial von Neumann algebra that embeds into an appropriate ultrapower of $\mathcal{R}_\infty$ must also embed into an ultrapower of $\mathcal{R}$.

In the case of $\mathcal{R}_\lambda$, there are states that lead to different universal theories. There is, however, a canonical state on $\mathcal{R}_\lambda$, which is known as the *Powers state* φ_λ. The following result also appears in [25].

Corollary 8.41 *For each* $\lambda \in (0, 1)$*, the universal theory of* $(\mathcal{R}_\lambda, \varphi_\lambda)$ *computes the halting problem.*

To prove the previous corollary, one uses the notion of the *centralizer* of a state. If φ is a state on M, the centralizer M_φ is the set of $x \in M$ for which $\varphi(xy) = \varphi(yx)$ for all $y \in M$. One then has that $\varphi|M_\varphi$ is a trace on M_φ. When $M = \mathcal{R}_\lambda$ and $\varphi = \varphi_\lambda$, we have that $M_\varphi = \mathcal{R}$. Moreover, in this setting, one can show that M_φ is effectively definable in M, whence one can use Theorem 8.35 above to encode the halting problem into the universal theory of $(\mathcal{R}_\lambda, \varphi_\lambda)$.

One can also use the ideas above to prove noncomputability results for certain C*-algebras. To state the results, in [18] the notion of the MEP for a structure M was introduced. Here, MEP stands for the statement that there is a satisfiable c.e. subtheory of $\mathrm{Th}(M)$ whose models embed into an ultrapower of M. This is slightly different from the setting discussed above in that we were considering arbitrary c.e. extensions of the theory of II_1 factors, but the same argument works with c.e. subsets of $\mathrm{Th}(\mathcal{R})$. Thus, the $\mathcal{R}$EP has a negative resolution.

Using this terminology, one can prove the following result from [18]. We will explain all of the undefined notions after the statement.

Corollary 8.42 *Suppose that (A, τ_A) is an infinite-dimensional, unital, simple, tracial* C*-*algebra with the uniform Dixmier property such that $N_{(A,\tau_A)}$ embeds into an ultrapower of $\mathcal{R}$. Then the AEP has a negative solution.*

First, a trace on a C*-algebra is simply a state that satisfies the trace property. The *uniform Dixmier property* is a certain (effectively axiomatizable) property that in particular implies that the algebra can admit at most one trace. Consequently, the state τ_A in the statement of the above theorem is unique. Next, by $N_{(A,\tau_A)}$ we mean the von Neumann algebra generated by A in the GNS representation corresponding to τ_A. The trace τ_A extends naturally to a faithful, normal trace on $N_{(A,\tau_A)}$.

There are many natural examples of C*-algebras that satisfy the above hypotheses. For example, the universal UHF algebra $\mathcal{Q}$ satisfies these assumptions, whence we see that the $\mathcal{Q}$EP has a negative solution. A well-known problem in C*-algebra theory known as the *Blackadar-Kirchberg Problem* asked if every *stably finite* C*-algebra embeds into an ultrapower of $\mathcal{Q}$.[4] While a negative solution to this problem follows quite immediately from a negative solution to the CEP, Corollary 8.42 in the case $A = \mathcal{Q}$ gives a Gödelian refutation of this problem as well.

While in the previous paragraph, the effective methods only gave a strengthening of a standard result, one can also use Corollary 8.42 above to prove new, purely operator algebraic results. Indeed, when applied to $A = \mathcal{Z}$, the so-called *Jiang-Su* algebra, an algebra to which the hypotheses of Corollary 8.42 apply, one can infer from the failure of the $\mathcal{Z}$EP that there are *stably projectionless*[5] C*-algebras which do not embed into an ultrapower of $\mathcal{Z}$. As far as we know, this result does not have a purely operator algebraic proof.

In the paper [24], Kirchberg introduced the notion of a C*-algebra with the QWEP. Kirchberg's QWEP Problem referred to above asked if all C*-algebras have the QWEP. As already mentioned, he proved that this problem is equivalent to the CEP, whence we now know that there are C*-algebras that do not have the QWEP (in fact, $C^*(\mathbb{F}_\infty)$ does not have the QWEP). In [26], Goldbring showed that the class of C*-algebras with the QWEP is an axiomatizable class. However, no concrete axioms for this class were offered. Instead, he showed that this class is closed under ultraproducts and ultraroots, which shows that the class is elementary by Theorem 2.14. Since this class is also closed under unions of chains, it is in fact an $\forall_2$-axiomatizable class. The reason no concrete axioms for this class were presented can now be explained by the following theorem of Arulseelan, Goldbring, and Hart [25], whose proof utilizes some of the same ideas as the proof of Corollary 8.42 above.

Corollary 8.43 *There is no c.e. theory T of* C**algebras, all of whose models have the QWEP, and which has an infinite-dimensional monotracial model whose unique trace is faithful. In particular, there is no c.e. set of axioms for the class of* C*-*algebras with the QWEP.*

---

[4] Here, a C*-algebra A is said to be *finite* if the identity is not Murray-von Neumann equivalent to any proper subprojection, while A is *stably finite* if $M_n(A)$ is finite for all $n \geq 1$.

[5] Projectionless C*-algebras were defined in the previous section. For a precise definition of what it means for a C*-algebra to be stably projectionless, see [18].

We remark that in a follow-up paper with Hart [27], the hypothesis of the first part of the corollary is significantly relaxed by requiring that T merely have some model which is not *subhomogeneous*.[6]

We now turn to the proof of Theorem 8.35 above. As already mentioned, the proof uses a landmark result in quantum complexity theory known as MIP* = RE. To state the result, we need to introduce some background material on nonlocal games. When n is a positive integer, set $[n] = \{1, \ldots, n\}$.

Definition 8.44 A *nonlocal game* $\mathfrak{G}$ with k questions and n answers is specified by two pieces of information.

1. A probability distribution π on $[k]^2$.
2. A decision predicate $D : [k]^2 \times [n]^2 \to \{0, 1\}$.

Here, one views the elements of $[k]$ as "questions" that two players, usually named Alice and Bob (the typical names used in quantum theory), can be asked. For $x, y \in [k]$, $\pi(x, y)$ describes the probability that Alice receives question x while Bob receives question y. Somehow (more on this later), Alice and Bob respond with "answers" a and b from $[n]$. Thinking of this process as a game, the decision predicate D then determines whether or not Alice and Bob won the game, that is, $D(x, y, a, b) = 1$ is interpreted as them winning the game, while $D(x, y, a, b) = 0$ is interpreted as them losing the game.

The term "nonlocal" refers to the fact that although Alice and Bob may cooperate, they are noncommunicating players who must decide on a strategy before they are sent far away from each other. Such a strategy is determined by the conditional probability distributions $p(a, b|x, y)$ that they respond with answers a and b when asked questions x and y respectively. There is a lot to say about these strategies, but the most important strategy for us to mention is that of a *quantum strategy*. Without getting into the details of how quantum theory works, let us say that a quantum strategy is determined by some "entangled" state ϕ that Alice and Bob share together with measurements that they can perform on their part of the state. Each of these measurements corresponds to one of the possible questions. Since measurement outcomes in quantum theory are probabilistic, the probability $p(a, b|x, y)$ is then determined by the probability that the measurements corresponding to questions x and y, when performed on the entangled state ϕ, lead to outcomes a and b respectively. The set of such quantum strategies is denoted $C_q(k, n)$.

Given a strategy p (quantum or otherwise), the expected value of their winning the game $\mathfrak{G}$ when playing according to strategy p, denoted $\mathrm{val}(\mathfrak{G}, p)$, is given by

$$\mathrm{val}(\mathfrak{G}, p) = \sum_{(x,y)\in[k]^2} \pi(x, y) \sum_{(a,b)\in[n]^2} D(x, y, a, b)p(a, b|x, y).$$

If Alice and Bob wish to maximize their probability of winning, they should consider the *quantum value* of the game given by

6 A C*-algebra is subhomogeneous if there is a bound on the dimension of its irreducible representations.

$$\mathrm{val}^*(\mathfrak{G}) = \sup_{p \in C_q(k,n)} \mathrm{val}(\mathfrak{G}, p).$$

A natural question arises: Is $\mathrm{val}^*(\mathfrak{G})$ computable from the data defining $\mathfrak{G}$? In other words, given $\mathfrak{G}$ and $\epsilon > 0$, can one compute $\mathrm{val}^*(\mathfrak{G})$ to within ϵ? A straightforward brute force search shows that $\mathrm{val}^*(\mathfrak{G})$ is left c.e. uniformly in $\mathfrak{G}$. The question then becomes: is $\mathrm{val}^*(\mathfrak{G})$ right c.e. uniformly in $\mathfrak{G}$? The main result of [20], known as MIP* = RE, is a resounding no!

Theorem 8.45 *There is an effective mapping from Turing machines* $\mathbf{M}$ *to nonlocal games* $\mathfrak{G}_{\mathbf{M}}$ *that satisfies the following.*

1. *If* $\mathbf{M}$ *halts, then* $\mathrm{val}^*(\mathfrak{G}_{\mathbf{M}}) = 1$.
2. *If* $\mathbf{M}$ *does not halt, then* $\mathrm{val}^*(\mathfrak{G}_{\mathbf{M}}) \leq \frac{1}{2}$.

Readers familiar with computational complexity theory will recognize the equation MIP* = RE as stating that two complexity classes coincide. The class RE is the class of computably enumerable languages,[7] while the class MIP* is the collection of those languages satisfying the conclusion of the previous theorem, with the Turing machine $\mathbf{M}$ being replaced by a finite string. The fact that one can give computable lower bounds to $\mathrm{val}^*(\mathfrak{G})$ shows that all languages in MIP* are computably enumerable. The previous theorem shows that the halting problem belongs to MIP*, whence all of RE belongs to MIP*.

In order to explain how Theorem 8.45 implies Theorem 8.35, we first define $C_{qa}(k,n)$ to be the closure (in $[0,1]^{k^2n^2}$) of $C_q(k,n)$. A result of Slofstra [28] shows that $C_q(k,n)$ is not closed, whence $C_{qa}(k,n)$ is a proper superset of $C_q(k,n)$. Note that $\mathrm{val}^*(\mathfrak{G})$ is unchanged if one takes the supremum over $C_{qa}(k,n)$ rather than $C_q(k,n)$.

We next define $p \in C_{qa}(k,n)$ to be *synchronous* if, for all $x \in [k]$, we have $p(a,b|x,x) = 0$ for all *distinct* $a, b \in [n]$. Informally, a strategy is synchronous if there is no chance that Alice and Bob respond with different answers when asked the same question. The set of such strategies is denoted by $C^s_{qa}(k,n)$, and we let $\mathrm{s}-\mathrm{val}(\mathfrak{G})$ denote the *synchronous quantum value* of $\mathfrak{G}$ given by

$$\mathrm{s}-\mathrm{val}^*(\mathfrak{G}) = \sup_{p \in C^s_{qa}(k,n)} \mathrm{val}(\mathfrak{G}, p).$$

It is important to note that Theorem 8.45 remains true if one replaces $\mathrm{val}^*(\mathfrak{G})$ with $\mathrm{s}-\mathrm{val}^*(\mathfrak{G})$.

The reason for passing to strategies in $C^s_{qa}(k,n)$ is that they admit a purely operator algebraic characterization, especially well suited for our purposes, as was shown in the following by Kim, Paulsen, and Schafhauser [29, Theorem 3.6]. Let τ denote the unique trace on $\mathcal{R}^{\mathcal{U}}$.

---

[7] The acronym RE stands for *recursively enumerable*. Recursively enumerable is synonymous with computably enumerable, but this terminology has fallen out of fashion with most computability theorists.

Theorem 8.46 *For $p \in [0,1]^{k^2n^2}$, we have that $p \in C_{qa}^s(k,n)$ if and only if there is a family $(e_j^x)_{x\in[k], j\in[n]}$ of projections in $\mathcal{R}^{\mathcal{U}}$ that satisfy the following two conditions.*

1. *For all $x \in [k]$, $\sum_{a\in[n]} e_a^x = 1$.*
2. *For all $x, y \in [k]$ and $a, b \in [n]$, $p(a,b|x,y) = \tau(e_a^x e_b^y)$.*

Let X denote the subset of the unit ball of $(\mathcal{R}^{\mathcal{U}})^{k^2n^2}$ consisting of such collections of projections, and let $\varphi_{\mathfrak{G}}(e)$ denote the atomic formula

$$\varphi_{\mathfrak{G}}(e) = \sum_{(x,y)\in[k]^2} \pi(x,y) \sum_{(a,b)\in[n]^2} D(x,y,a,b)\tau(e_a^x e_b^y).$$

We then have that

$$\mathrm{s}-\mathrm{val}^*(\mathfrak{G}) = \sup_{e\in X} \varphi_{\mathfrak{G}}^{\mathcal{R}^{\mathcal{U}}}(e).$$

It turns out that X is a definable subset of $\mathcal{R}^{\mathcal{U}}$ (as defined in Sect. 2.4.2). In fact, the verification that X is a definable subset of $\mathcal{R}^{\mathcal{U}}$ was performed by Kim, Paulsen, and Schafhauser themselves in [29, Lemma 3.5] (although they did not use the model-theoretic terminology of definable set). Consequently, there is a sentence $\psi_{\mathfrak{G}}$ such that $\mathrm{s}-\mathrm{val}^*(\mathfrak{G}) = \psi_{\mathfrak{G}}^{\mathcal{R}^{\mathcal{U}}}$. Moreover, as mentioned in Remark 2.21, by carefully analyzing how one converts expressions obtained by quantifying over definable sets with actual formulas, one can show that the mapping from $\mathfrak{G}$ to $\psi_{\mathfrak{G}}$ is an effective mapping and, moreover, these sentences are *universal* sentences. Consequently, if the universal theory of $\mathcal{R}$ were computable, one would be able to effectively approximate $\psi_{\mathfrak{G}}^{\mathcal{R}^{\mathcal{U}}}$ and thus effectively approximate $\mathrm{s}-\mathrm{val}^*(\mathfrak{G})$. By Theorem 8.45 above (with val^* replaced by $\mathrm{s}-\mathrm{val}^*$), we would be able to decide the halting problem.

It is possible to give a reformulation of the fact that the universal theory of $\mathcal{R}$ is not computable in purely operator algebraic terms. Given positive integers n and d, let $(m_j)_{j\le L(n,d)}$ denote a one-to-one enumeration of all $*$-mononomials in the variables $x_1, \ldots, x_n$ of total degree at most d. We then consider the map $\mu_{n,d} : \mathcal{R}_1^n \to \mathbb{D}^L$ given by $\mu_{n,d}(\mathbf{a}) = (\tau(m_i(\mathbf{a})))_{i\in[L(n,d)]}$. (Here, $\mathbb{D}$ is the complex unit disk.)

We let $X(n,d)$ denote the range of $\mu_{n,d}$ and $X(n,d,p)$ be the image of the unit ball of $M_p(\mathbb{C})$ under $\mu_{n,d}$. Notice that $\bigcup_{p\in\mathbb{N}} X(n,d,p)$ is dense in $X(n,d)$.

Theorem 8.47 *The following statements are equivalent.*

1. *The universal theory of $\mathcal{R}$ is computable.*
2. *There is a computable function $F : \mathbb{N}^3 \to \mathbb{N}$ such that, for every $n, d, k \in \mathbb{N}$, $X(n,d,F(n,d,k))$ is $\frac{1}{k}$-dense in $X(n,d)$.*

We end this section with the obvious question posed by Theorem 8.35.

Question 8.48 If n is a positive integer, does the Π_n theory of $\mathcal{R}$ compute $\emptyset^{(n)}$?

References

1. Itaï Ben Yaacov, Alexander Berenstein, C. Ward Henson, and Alexander Usvyatsov. Model theory for metric structures. *London Mathematical Society Lecture Note Series*, 350:315, 2008.
2. Caleb Camrud, Isaac Goldbring, and Timothy H. McNicholl. On the complexity of the theory of a computably presented metric structure. *Archive for Mathematical Logic*, 62(7-8):1111–1129, 2023.
3. C. J. Ash and J. Knight. *Computable Structures and the Hyperarithmetical Hierarchy*, volume 144 of *Studies in Logic and the Foundations of Mathematics*. North-Holland Publishing Co., Amsterdam, 2000.
4. Caleb M. H. Camrud and Timothy H. McNicholl. Hyperarithmetic numerals. In *Twenty Years of Theoretical and Practical Synergies*, volume 14773 of *Lecture Notes in Comput. Sci.*, pages 277–284. Springer, Cham, 2024.
5. Isaac Goldbring and Bradd Hart. Operator algebras with hyperarithmetic theory. *Journal of Logic and Computation*, 31(2):612–629, 2021.
6. Joram Hirschfeld and William H. Wheeler. *Forcing, arithmetic, division rings*, volume 454 of *Lecture Notes in Mathematics*. Springer, Berlin-New York, 1975.
7. Isaac Goldbring. Enforceable operator algebras. *Journal of the Institute of Mathematics of Jussieu*, 20(1):31–63, 2021.
8. Isaac Goldbring, Bradd Hart, and Thomas Sinclair. The theory of tracial von Neumann algebras does not have a model companion. *The Journal of Symbolic Logic*, 78(3):1000–1004, 2013.
9. Christopher James Eagle, Ilijas Farah, Eberhard Kirchberg, and Alessandro Vignati. Quantifier elimination in C*-algebras. *International Mathematics Research Notices*, 2017(24):7580–7606, 2017.
10. Christopher J. Eagle and Alessandro Vignati. Saturation and elementary equivalence of C*-algebras. *Journal of Functional Analysis*, 269(8):2631–2664, 2015.
11. Christopher J. Eagle, Isaac Goldbring, and Alessandro Vignati. The pseudoarc is a co-existentially closed continuum. *Topology and its Applications*, 207:1–9, 2016.
12. R. H. Bing. The pseudo-arc. In *Summary of Lectures and Seminars, Summer Institute on Set Theoretic Topology, Madison, Wisconsin*, pages 72–75, 1955.
13. John G. Hocking and Gail S. Young. *Topology*. Dover Publications Inc., New York, second edition, 1988.
14. Itaï Ben Yaacov and C. W̃ard Henson. Generic orbits and type isolation in the Gurarij space. *Fund. Math.*, 237:47–82, 2017.
15. Isaac Goldbring and Thomas Sinclair. On Kirchberg's embedding problem. *Journal of Functional Analysis*, 269(1):155–198, 2015.
16. Isaac Goldbring and Martino Lupini. Model-theoretic aspects of the Gurarij operator system. *Israel Journal of Mathematics*, 226(1):87–118, 2018.
17. Martino Lupini. Uniqueness, universality, and homogeneity of the noncommutative Gurarij space. *Advances in Mathematics*, 298:286–324, 2016.
18. Isaac Goldbring and Bradd Hart. The universal theory of the hyperfinite II_1 factor is not computable. *Bull. Symb. Log.*, 30(2):181–198, 2024.
19. Isaac Goldbring and Bradd Hart. Computability and the Connes embedding problem. *Bulletin of Symbolic Logic*, 22(2):238–248, 2016.
20. Zhengfeng Ji, Anand Natarajan, Thomas Vidick, John Wright, and Henry Yuen. MIP*=RE. *Communications of the ACM*, 64(11):131–138, 2021.
21. Tobias Fritz. Tsirelson's problem and Kirchberg's conjecture. *Reviews in Mathematical Physics*, 24(05):1250012, 2012.
22. Marius Junge, Miguel Navascues, Carlos Palazuelos, David Perez-Garcia, Volkher B Scholz, and Reinhard F Werner. Connes' embedding problem and Tsirelson's problem. *Journal of Mathematical Physics*, 52(1), 2011.
23. Narutaka Ozawa. About the QWEP conjecture. *International Journal of Mathematics*, 15(05):501–530, 2004.

24. Eberhard Kirchberg. On non-semisplit extensions, tensor products and exactness of group C*-algebras. *Inventiones mathematicae*, 112:449–489, 1993.
25. Jananan Arulseelan, Isaac Goldbring, and Bradd Hart. The undecidability of having the QWEP. *Journal of Operator Theory*, 92:349–362, 2024.
26. Isaac Goldbring. Model theory and the QWEP conjecture. *Illinois Journal of Mathematics*, 59(3):795–799, 2015.
27. Isaac Goldbring and Bradd Hart. On Tsirelson pairs of C*-algebras. *Reviews in Mathematical Physics*, 35(08):2350016, 2023.
28. William Slofstra. The set of quantum correlations is not closed. *Forum of Mathematics, Pi*, 7, 2019.
29. Se-Jin Kim, Vern Paulsen, and Christopher Schafhauser. A synchronous game for binary constraint systems. *Journal of Mathematical Physics*, 59(3), 2018.

Appendix A
Collected Open Questions

In this appendix, we collect together the questions and problems posed in prior chapters, and we pose some additional avenues for investigation as well. Our questions are organized according to the kinds of structures we have studied.

A.1 Metric Spaces

The results in Chap. 4 concern the computable categoricity of many common kinds of metric spaces, namely subsets of Euclidean space, Cantor space, and Baire space. They also address one example of an unusual space, the Urysohn space. Classical point-set topology provides many other examples of exotic metric spaces that have not yet been investigated. For example, we pose the following question.

Question A.1 Is the pseudoarc $\mathbb{P}$ (see Example 8.32) computably categorical as a metric space?

In Chap. 4, some examples of metric spaces that are not computably categorical are presented. However, other than Theorem 4.29, not much is known about relative categoricity and degrees of categoricity. We are thus led to raise the following questions.

Question A.2 Does the metric space $C([0, 1]; \mathbb{R})$ have a degree of categoricity and, if so, what is it?

Question A.3 Does the metric space $C(2^{\mathbb{N}}; \mathbb{R})$ have a degree of categoricity and, if so, what is it?

Question A.4 What are the degrees of categoricity of compact metric spaces?

There are many results on index sets for classes of metric spaces, but there are many more that could be explored. We list a few of them here.

J. N. Y. Franklin et al., *Effective Metric Structure Theory*, Springer Monographs in Mathematics, https://doi.org/10.1007/978-3-032-13322-9

Question A.5 What is the complexity of the index set of the metric space $C([0, 1]; \mathbb{R})$?

Question A.6 What is the complexity of the index set of the metric space $C(2^{\mathbb{N}}; \mathbb{R})$?

Question A.7 If X is an infinite computably presentable metric space, does it follow that the index set of $\{M \mid X \hookrightarrow M\}$ is Σ^1_1-complete?

A.2 Banach Spaces

A.2.1 C-Spaces

$C([0, 1]; \mathbb{F})$

With regards to $C([0, 1]; \mathbb{F})$, the main open questions regard its relative categoricity and its index set.

Question A.8 Is the bound in Theorem 5.9 optimal? In particular, does $C([0, 1]; \mathbb{F})$ have a degree of categoricity, and if so, what is it?

Question A.9 Is the upper bound of Σ^0_7 for the index set of $C([0, 1]; \mathbb{R})$ sharp? In other words, is this index set Σ^0_7-complete?

$C(2^{\mathbb{N}}; \mathbb{F})$

For each $\mathbb{F} \in \{\mathbb{R}, \mathbb{C}\}$, the relative categoricity of $C(2^{\mathbb{N}}; \mathbb{F})$ has not been investigated, nor has its index set.

Problem A.10 For each $\mathbb{F} \in \{\mathbb{R}, \mathbb{C}\}$, find the degree of categoricity (if there is one) of $C(2^{\mathbb{N}}; \mathbb{F})$.

Problem A.11 For each $\mathbb{F} \in \{\mathbb{R}, \mathbb{C}\}$, find the complexity of the index set of $C(2^{\mathbb{N}}; \mathbb{F})$.

Other *C*-Spaces

C-spaces for other kinds of compact metrizable spaces are still open areas of inquiry. We list three possible questions here.

Question A.12 If α is a countable successor ordinal for which $C(\alpha; \mathbb{F})$ is computably presentable, does it follow that α is computable?

Problem A.13 Explore the computable structure theory of $C(\alpha; \mathbb{F})$ for $\alpha < \omega_1^{ck}$ a successor ordinal.

Problem A.14 Explore the computable structure theory of $C(\mathbb{P}; \mathbb{F})$, where $\mathbb{P}$ is the pseudoarc.

Finally, the classification of computably presentable C-spaces is still very much open.

Problem A.15 For each $\mathbb{F} \in \{\mathbb{R}, \mathbb{C}\}$, classify the compact separable metric spaces X such that $C(X; \mathbb{F})$ is computably presentable.

A.2.2 Lebesgue Spaces

The fundamental open question regarding Lebesgue spaces concerns their computable presentability, which has proven to be surprisingly difficult. For example, the results in Chap. 6 lead us to the following conjecture.

Conjecture A.16 *If $1 \leq p < 2$ is right c.e., then $L^p[0, 1]$ is computably presentable.*

If Conjecture A.16 is correct, then there will be a wealth of other questions to answer, such as those regarding computable categoricity. More generally, we have the following classification problem.

Problem A.17 Classify the computably presentable L^p spaces for $1 \leq p < 2$.

Finally, there is the index set for the classification of Lebesgue spaces.

Problem A.18 Determine the complexity of the index set of all Lebesgue spaces.

In Chap. 6, we conjectured that this set is Π^0_3-complete.

A.2.3 Exotic Banach Spaces

The specific Banach spaces investigated so far might be viewed as rather ordinary, even though they lead to some very difficult problems. For those seeking more unusual terrain, we offer the following questions.

Question A.19 Does the Gurarij Banach space $\mathbb{G}$, noncommutative Gurarij space $\mathbb{NG}$, or Gurarij operator system $\mathbb{GS}$ (see Example 8.33) have a computable presentation?

Question A.20 Is the Tsirelson space computably presentable?

Question A.21 Is James' Banach space computably presentable?

A.2.4 *Banach Spaces in General*

Chapters 5 and 6 focused on particular and fairly concrete classes of Banach spaces. With regards to Banach spaces in general, a tremendous amount is unknown. We list a selection of problems here.

Question A.22 If a *complex* Banach space is computably categorical as a metric space, does it follow that it is computably categorical? (Recall that in Proposition 4.8, we showed that the question has a positive answer when the Banach space is real.)

Question A.23 Is there a Banach space with intermediate computable dimension?

Question A.24 What are the possible degrees of categoricity of Banach spaces?

A theory is *n-decidable* if its Π_n theory is decidable.

Question A.25 Fix a positive integer n. Is there an $(n+1)$-decidable Banach space that is not n-decidable?

Question A.26 For each $n \in \mathbb{N}$, is there a Banach space that is $\mathbf{0}^{(\mathbf{n+1})}$-categorical but not $\mathbf{0}^{(\mathbf{n})}$-categorical?

Question A.27 Is there a computably presentable Banach space whose dual is separable but not computably presentable?

Problem A.28 Determine the complexity of the set of indices of computable presentations of strictly convex Banach spaces.

Problem A.29 Determine the complexity of the set of indices of computable presentations of reflexive Banach spaces.

Conjecture A.30 *The set of all indices of computably categorical Banach spaces is* Σ^1_1.

A.3 Operator Algebras

A.3.1 *Abelian Unital Operator Algebras*

By the Gelfand Theorem, the unital abelian C*algebras are precisely those of the form $C^*(X)$. The general computability theory of these algebras is already resolved by the results in Chap. 7. However, there are particular cases that are still uninvestigated.

Question A.31 Does $C^*([0, 1])$ have a degree of categoricity and, if so, what is it?

Problem A.32 Explore the computable structure theory of $C^*(\mathbb{P})$, where $\mathbb{P}$ is the pseudoarc.

Problem A.33 Explore the computable structure theory of $C^*(\alpha)$, where $\alpha < \omega_1^{ck}$ is a successor ordinal.

At present, we do not have natural but significant examples of operator algebras with decidable theories. The theory of $C^*(2^{\mathbb{N}})$ is one of the best candidates.

Question A.34 Is $\mathrm{Th}(C^*(2^{\mathbb{N}}))$ computable?

A.3.2 Nonabelian and Possibly Nonunital Operator Algebras

As pointed out in Chap. 5, the unit of a unital abelian C*-algebra $\mathcal{A}$ is a computable point of all computable presentations of $\mathcal{A}$. However, as shown in [1], this principle does not hold uniformly. In the nonabelian case, this rather simple question remains unresolved.

Question A.35 Is there a computable presentation $A^{\#}$ of a C*-algebra A for which $\mathbf{1}_A$ is not a computable point?

In a forthcoming paper [2], Eagle, Goldbring, McNicholl, and R. Miller show that the unit is a computable point of any computable presentation of a UHF algebra. With regards to Question A.35, this is the best partial result to date. There is much to be done that concerns the complexity of theories for various operator algebras. Because of the role of $\mathcal{R}$ in the theory of tracial von Neumann algebras, we believe the following is the most significant of these questions.

Question A.36 If n is a positive integer, does the Π_n theory of $\mathcal{R}$ compute $\emptyset^{(n)}$?

A.3.3 von Neumann Algebras

It is surprisingly difficult to demonstrate the computable presentability of a von Neumann algebra. Thus, it is not surprising that their computable categoricity remains unexplored.

Question A.37 Are any of the computably presentable tracial von Neumann algebras discussed herein computably categorical?

A.3.4 Operator Algebras in General

Chapter 7 left open two questions regarding the word problem for C*algebras.

Question A.38 Is there a presentation $\mathcal{A}^{\dagger}$ of a C*-algebra $\mathcal{A}$ such that WP($\mathcal{A}^{\dagger}$) is computably closed and yet $\mathcal{A}^{\dagger}$ is not computable?

Question A.39 If $G^{\dagger}$ is computable, must WP($C^*(G^{\dagger})$) and WP($C^*_r(G^{\dagger})$) be computably closed?

There is much regarding the general computability theory of operator algebras that remains unexplored.

Question A.40 Is there an operator algebra with intermediate computable dimension?

Question A.41 What are the possible degrees of categoricity of operator algebras?

Question A.42 For each $n \in \mathbb{N}$, is there an operator algebra that is $\mathbf{0^{(n+1)}}$-categorical but not $\mathbf{0^{(n)}}$-categorical?

Question A.43 Fix a positive integer n. Is there an $(n+1)$-decidable operator algebra that is not n-decidable?

A.4 Metric Structures in General

Question A.44 Is there a language L with finitely many predicate symbols and a computably presentable L-structure $\mathcal{M}$ that achieves all of the upper bounds in Theorem 8.1?

Again, we conjecture that the theory of $\mathcal{R}$ provides a positive answer.

A.5 Concluding Remarks

A good part of our enthusiasm for this subject, which prompted us to write this book, stems from its ability to blend methods from multiple disparate fields to achieve solutions of the problems it poses. We believe the resolution of the questions and problems we have set forth here will lead to new discoveries of deep and exciting connections between analysis, model theory, and computability theory, and we hope that the reader will feel inspired to engage in their pursuit.

Appendix B
Computability Theory Background

The purpose of this appendix is to provide the reader with the fundamental notions of computability theory. While several fine volumes on the subject exist, we hope to provide a quick foundation that will allow the reader unfamiliar with computability to dive straight into the rest of the book or provide a reference for a reader who is somewhat familiar with the area. Should the reader wish for a complete reference, we recommend Cooper [3] or Odifreddi [4, 5]; a quicker introduction, oriented towards algorithmic randomness, can be found in Chap. 2 of [6].

B.1 Notation

Our notation is standard as of the time of this writing. We denote natural numbers by lowercase Latin letters such as k and n, and subsets of $\mathbb{N}$ will be denoted by uppercase Latin letters such as A and B.

B.2 Basic Computability

Modern computability theory originated in the 1930s. Different methods of formalizing the notion of an effective function, such as the λ-calculus, μ-recursion, register machines, and Turing machines were described. However, it was found that all of these formalizations described the same class of functions, and these days it is typical not to fix a formalization in circumstances such as ours. Instead, we will simply say that we are considering the kinds of calculations that a computer could carry out in a finite but unbounded amount of time given a finite but unbounded amount of memory.

J. N. Y. Franklin et al., *Effective Metric Structure Theory*, Springer Monographs in Mathematics, https://doi.org/10.1007/978-3-032-13322-9

B.2.1 Partially Computable Functions and Computably Enumerable Sets

The most fundamental type of function we will consider is a *partial computable* function, which is exactly the type of calculation described above. We may think of such a function as one that can be computed by an idealized computer program that accepts any natural number as input and outputs natural numbers. The word *partial* is an important one: those of us with programming experience are well aware that a computer program may not produce an output for each possible input. However, if it does, it must produce this output (*halt*) at a finite point, or *stage*. If a partial computable function halts on each possible input, we call it simply a *computable function*. Partial computable functions are usually denoted by lowercase Greek letters like φ or ψ.

At this point, we can make some important observations about these functions. First and foremost, there are only countably many partial computable functions, and we can list them in a computable way. The first part of this statement follows by observing that the idealized computer programs we have discussed can be viewed as finite sequences of symbols from a finite alphabet, and there are only countably many such sequences. The second part follows from the fact that we can list these computer programs in an effective way, for instance, length-lexicographically. This allows us to write a computable list of the partial computable functions as $\varphi_0, \varphi_1, \varphi_2, \ldots$.

Second, while there are only countably many computable functions by the argument given above, we cannot list them computably because we cannot effectively determine which partial computable functions will halt on each possible input.

Finally, we note that many useful sets are computably bijective to $\mathbb{N}$ such as $\mathbb{N} \times \mathbb{N}$ and the rationals $\mathbb{Q}$; we write the computable pairing function from $\mathbb{N}^2$ to $\mathbb{N}$ as $\langle \cdot, \cdot \rangle$. This lets us see that, while we stated above that a partial computable function must have $\mathbb{N}$ as its domain and codomain, we could instead interpret such a function as having such a set as its domain or codomain.

Now we can introduce some useful notation. If the function φ_e halts on input k, we say that $\varphi_e(k)$ *converges* and write $\varphi_e(k)\downarrow$. Otherwise, we say that $\varphi_e(k)$ *diverges* and write $\varphi_e(k)\uparrow$. We may also write $\varphi_e(k)[s]\downarrow= m$ if $\varphi_e(k)$ halts at or before stage s with output m (we may see $\varphi_{e,s}(k)\downarrow= m$ in the literature, but this is less usual at the time of this writing).

Now we turn our attention to the kinds of sets that we can generate with partial computable functions. The most basic is a *computably enumerable (c.e.)* set, which is a set that is the domain of a partial computable function. Just as the partial computable functions can be listed $\varphi_0, \varphi_1, \varphi_2, \ldots$, we can list the computably enumerable sets $W_0, W_1, W_2, \ldots$. We can gain intuition about the terminology we use here. To enumerate W_e, we begin by running φ_e on input 0 for one stage; if it halts, we enumerate 0 into W_e. Then we iterate this process. On our k^{th} iteration, we run φ_e on all of the inputs $0, \ldots, k-1$ for k stages; if it halts on any of these inputs for the first time at that stage, we enumerate those inputs into W_e. We use similar notation for approximations to c.e. sets as we have seen for partial computable functions: the set of elements that have been iterated into W_e by stage s is denoted $W_e[s]$ (less frequently, $W_{e,s}$).

We mention that there are many equivalent ways of characterizing the c.e. sets in addition to the one we used above; we list several here. One final characterization will be given in two pages, when we discuss the arithmetic hierarchy.

Theorem B.1 *The following are equivalent for a set* $A \subseteq \mathbb{N}$*:*

- *A is c.e.*
- *A is the domain of a partial computable function.*
- *A is the range of a partial computable function.*
- *A is the range of a total computable function or* $A = \emptyset$*.*

We also present one more characterization of a partial computable function in terms of c.e. sets. The *graph* of a function φ is the set $\{(k, m) \ : \varphi(k) = m\}$, and a function is partial computable exactly when its graph is c.e.

We observe that there are many effective ways to list the partial computable functions or to enumerate the computably enumerable sets. However, just as the formalization we choose for computability theory does not matter for our purposes, the precise enumeration of the partial computable functions or computably enumerable sets does not matter.

We note that, just as we have defined computable functions, we can define a *computable* set as one that is both c.e. and has a c.e. complement. The elements of such a set can be computably identified: to determine whether k belongs to the set, we begin enumerating both the set and its complement and wait to see which one k belongs to.

We will also frequently use these concepts to discuss sequences of sets. For instance, given an effective listing of the c.e. sets $W_0, W_1, W_2, \ldots$ as above, we can define a *sequence of uniformly c.e. sets* using a computable function f to pick out the indices in our listing: $f(i)$ will be the index of the ith c.e. set in our sequence.

In the next subsection, it will be useful to view a subset of $\mathbb{N}$ as an infinite binary sequence: if k belongs to our subset, then the $(k+1)^{\text{st}}$ bit of the sequence is a 1, and otherwise the $(k+1)^{\text{st}}$ bit is a 0.

B.2.2 Turing Functionals and Turing Degrees

Now that we have a working knowledge of the fundamental types of functions and sets in computability theory, we turn our attention to our primary method for comparing the computational strength of two sets. In order to do this, we need the notion of oracle computation. A *Turing functional* takes a natural number k as input, like a partial computable function, but also allows the use of an *oracle*, or an infinite binary sequence whose initial segments can be used as information in a computation. These Turing functionals can be thought of as having the same computational ability as a partial computable function plus the ability to ask and answer questions of the form "Does k belong to the oracle?", so they can be listed effectively as well.[1] In fact, most of our notation is analogous to that for partial computable functions. We typically use capital Greek letters for the functionals themselves and write our effective list as $\Phi_0, \Phi_1, \Phi_2, \ldots$. When we use an infinite binary sequence A as an oracle for a functional Φ, we will write Φ^A. When a functional Φ with oracle A halts on input k within s steps, we write $\Phi^A(k)[s]\downarrow$.

It is important to note that, even though a Turing functional Φ may have access to an infinite binary string as its oracle, it will never access more than finitely many of its bits during any given computation. This means that we will sometimes use a finite binary string as an oracle for a given computation. For instance, we may write $\Phi^\tau(k)[s]$ instead of $\Phi^A(k)[s]$ if τ is an initial segment of A and all of the bits of A accessed during the first s stages of our computation of $\Phi^A(k)$ are in τ.

Our primary use of Turing functionals will be to compare the computational strengths of two sets. We say that a set $A \subseteq \mathbb{N}$ is *computable from* a set $B \subseteq \mathbb{N}$ if there is a Turing functional that, given B as an oracle, computes the characteristic function of A. We may also say that A is *Turing reducible to* B, or *computable relative to* B, and we write this as $A \leq_T B$. If $A \leq_T B$ and $B \leq_T A$, we say that A and B are *Turing equivalent* and write $A \equiv_T B$. The set $\{B : A \equiv_T B\}$ is called the *Turing degree* of A and will generally be written as $\mathbf{a}$. We may also write $\mathbf{a} \leq_T \mathbf{b}$ to indicate that any element of $\mathbf{a}$ is Turing reducible to any element of $\mathbf{b}$.

We can also define new sets and degrees using Turing functionals: for any set A, we can define the *jump* of A, written A', to be $\{k : \Phi_k^A(k)\downarrow\}$. This lends itself to a natural partial order. The base of this partial order is the Turing degree consisting of the computable sets, which we call $\mathbf{0}$ (note that for every set A, $\emptyset \leq_T A$). Then the first step of this hierarchy is $\emptyset' = \{k : \Phi_k^\emptyset(k)\downarrow\}$. This is called the halting set and formalizes the Entscheidungsproblem, and its Turing degree $\mathbf{0}'$ appears frequently in the context of this book.

We can create a linear order within this partial order by iterating the jump operator, starting with $\mathbf{0}$: $\mathbf{0} <_T \mathbf{0}' <_T \mathbf{0}'' <_T \ldots$. As is standard, we write the nth iteration of the Turing jump of A as $A^{(n)}$ for readability.

B.2.3 The Arithmetic Hierarchy

We now discuss the arithmetic hierarchy and explain its relationship to the linear order described at the end of the previous subsection. We begin by noting that a subset of $\mathbb{N}$ can often be characterized by the number of unbounded quantifiers required to state its definition using only computable relations. The "simplest" subsets of $\mathbb{N}$ can be defined with, at the most, bounded quantifiers. For instance, the empty set can be defined without any quantifiers, and the set of prime numbers can be defined using only bounded quantifiers:

[1] Note that, while the oracle is an infinite binary sequence, we are thinking of it as a set as per the previous subsection.

$$\{k \in \mathbb{N} : (\forall i < k)[i | k \Rightarrow i = 1]\}.$$

In general, we say that a set is Σ_n if its definition can be stated with n alternations of quantifiers, starting with an existential quantifier, applied to a computable relation R of the appropriate arity:

$$(\exists x_1)(\forall x_2) \ldots (Q_n x_n) R(x_1, x_2, \ldots, x_n, y).$$

Here, Q_n will be existential if n is odd and universal if n is even.

Similarly, a set is Π_n if its definition can be stated with n alternations of quantifiers, starting with an universal quantifier, applied to a computable relation R of the appropriate arity:

$$(\forall x_1)(\exists x_2) \ldots (Q_n x_n) R(x_1, x_2, \ldots, x_n, y).$$

Here, Q_n will be universal if n is odd and existential if n is even. A Π_n set is thus the complement of a Σ_n set and vice versa.

We note that in both scenarios above, each quantifier written above could be construed as a block of identical quantifiers condensed into a single quantifier through use of the computable pairing function mentioned earlier.

The first example we propose is that of $\emptyset'$, which is Σ_1:

$$\emptyset' = \{k : (\exists s)[\Phi_k^{\emptyset}(k)[s] \downarrow]\}.$$

The set of e such that W_e is infinite, on the other hand, is Π_2:

$$\{e : (\forall k)(\exists i)(\exists s)[i > k \; \& \; \varphi_e(i)[s] \downarrow]\}.$$

Finally, the Δ_n sets are precisely those that are both Σ_n and Π_n. The Σ_n, Π_n, and Δ_n sets form the *arithmetic hierarchy*.[2] Often, we will simply write, for instance, Σ_n for the class of Σ_n sets. We make some quick observations about this hierarchy:

- The class of computable sets is $\Sigma_0 = \Pi_0 = \Delta_0 = \Delta_1$.
- A set is c.e. exactly when it is Σ_1.
- For $n \geq 1$, a set is Δ_n exactly when it is Turing computable from $\emptyset^{(n-1)}$.
- The arithmetic hierarchy does not collapse: for $n \geq 1$, we have $\Delta_n \subsetneq \Sigma_n$ and $\Delta_n \subsetneq \Pi_n$, and for all n, we have $\Sigma_n \subsetneq \Sigma_{n+1}$ and $\Pi_n \subsetneq \Pi_{n+1}$.

We may also define the complexity of a set in terms of *differences* of Σ_n sets for some fixed n. At the simplest, we say that a set A is *m-c.e.* if we have a computable function $f : \mathbb{N} \times \mathbb{N} \to \{0, 1\}$ such that:

[2] A superscript 0 is often included in this notation: Σ_n^0, Π_n^0, and Δ_n^0. The reason for this will be made clear in Sect. B.3.

- $f(k, 0) = 0$ for all k;
- $f(k, s) \neq f(k, s + 1)$ for at most m values of s; and
- the characteristic function of A is $\lim_{s\to\infty} f(k, s)$.

In short, we can "change our minds" about whether k belongs to A up to m times.

The above definition of an m-c.e. set could also be generalized to m-Σ_n sets: A set is m-Σ_n if it satisfies the above definition for an f that is computable from $\emptyset^{(n-1)}$.

We make two final observations about the terminology for this kind of set. The first is that, while the complement of a Σ_n set is a Π_n set, we do not have a clear way of denoting the complement in this context and thus call the complement of an m-Σ_n set a co-m-Σ_n set. We further note that a 2-c.e. set is often called a d-c.e. set (in this case, the d stands for "difference"), and similarly a 2-Σ_n set is called a d-Σ_n set.

To introduce the last notions of this section, we must define another reducibility. We say that a set A is *many-one reducible (m-reducible) to* a set B if there is a computable function f such that $x \in A$ if and only if $f(x) \in B$.

We say that a set A is Σ_n*-hard* if every Σ_n set B is m-reducible to A, and we say that A is Σ_n*-complete* if it is Σ_n-hard and Σ_n. It can be proven that $\emptyset^{(n)}$ is Σ_n-complete for every n. The corresponding notions for Π_n and other such classes are defined analogously.

B.2.4 Classes of Effectively Approximable Reals and Real-Valued Functions

We now turn our attention to ways to describe certain types of reals that will be useful. The reader will be able to see parallels with the types of sets we discussed in Sect. B.2 of this appendix.

Here, for $A \subseteq \mathbb{N}$, we can define $0.A$ as shorthand for $\sum_{n\in A} 2^{-(n+1)}$.

First, we would like to define a *computable* real number to be one that is effectively approximable to any degree of precision desired. This is, in fact, the definition most commonly used in practice:

Theorem B.2 ([6]) *The following are equivalent for a real number r:*

1. *r is computable.*
2. *There is a computable sequence of rationals $(q_i)_{i\in\mathbb{N}}$ converging to r such that for all i, we have $|r - q_i| < 2^{-i}$. Such a sequence is called a* Cauchy name *for r.*
3. *There is a computable sequence of rationals $(q_i)_{i\in\mathbb{N}}$ converging to r and a computable function f such that for all i, $|r - q_{f(i)}| < 2^{-i}$.*

It is now reasonable to wonder whether there is a sort of real that is analogous to a c.e. set, just as computable reals are analogous to computable sets. Indeed, we can talk about *left-c.e.* reals, also known as *lower semicomputable* reals; the characterizations below are those most commonly used in practice.

Theorem B.3 ([6]) *The following are equivalent for a real number r:*

1. *r is left-c.e.*
2. *There is a computable sequence of rationals $(q_i)_{i\in\mathbb{N}}$ converging to r such that $q_0 < q_1 < q_2 < \ldots$.*
3. *There is a computable sequence of positive rationals $(q_i)_{i\in\mathbb{N}}$ such that $r = \sum q_i$.*

It is also equivalent to say that the left Dedekind cut of r is c.e. It is natural to wonder how far this equivalence extends, namely, if a real $r = 0.A$ is left c.e., is the corresponding set A c.e. (and vice versa)? The answer is not so simple: there exist left-c.e. reals $0.A$ such that A is not a c.e. set.

We note that there are also *right-c.e.* reals, also known as *upper semicomputable* reals; we will simply note that a real is right c.e. if there is a computable sequence of rationals $(q_i)_{i\in\mathbb{N}}$ converging to r such that $q_0 > q_1 > q_2 > \ldots$. This is equivalent to saying that the right Dedekind cut of r is c.e. We note that a real is computable if and only if it is left c.e. and right c.e.

Suppose $(r_n)_{n\in\mathbb{N}}$ is a sequence of reals. We say that the sequence $(r_n)_{n\in\mathbb{N}}$ is *computable* if r_n is computable uniformly in n, that is, from $n \in \mathbb{N}$ it is possible to compute an index of a Turing machine that computes r_n. We also say that the sequence $(r_n)_{n\in\mathbb{N}}$ is *lower semicomputable* if r_n is left c.e. uniformly in n and *upper semicomputable* if r_n is right c.e. uniformly in n.

B.3 Higher Computability

It will also be necessary for us to be familiar with computability theory "beyond $\mathbb{N}$" when we discuss the infinitary results in Chap. 8. We begin with a discussion of computable ordinals. A reader wishing for more details should consult [7].

Perhaps the most intuitive way to think of the computable ordinals is via computable well orderings: a well ordering $(A, <_A)$ is computable if A is a subset of $\mathbb{N}$ and the relation $<_A$ is defined in a computable way. For instance, the well ordering

$$0, 2, 4, \ldots, 1, 3, 5, \ldots$$

is computable. We can then say that an ordinal α is computable if some computable well ordering has order type α. For instance, all natural numbers and ω are computable ordinals, and the well ordering defined above shows that $\omega + \omega$ is also a computable ordinal. We can observe that the computable ordinals form an initial segment of the ordinals and that the successor of any computable ordinal is computable.

The supremum of all computable ordinals is the first noncomputable ordinal and is denoted ω_1^{ck}.[3] Note that ω_1^{ck} is countable.

We now turn to another way of defining the computable ordinals: ordinal notations. A system of notations for ordinals consists of a set $\mathcal{O}$, a function $|\cdot|_{\mathcal{O}}$ that maps

---

[3] Here, ck stands for Church-Kleene.

each element of $\mathcal{O}$ to an ordinal, and a partial order $<_{\mathcal{O}}$ on $\mathcal{O}$. This system is defined inductively as follows:

- The notation for the ordinal 0 is 1, so $|1|_{\mathcal{O}} = 0$.
- We now consider a successor ordinal $\alpha + 1$. If $|a|_{\mathcal{O}} = \alpha$, then $|2^a|_{\mathcal{O}} = \alpha + 1$. We declare that $b <_{\mathcal{O}} 2^a$ if either $b <_{\mathcal{O}} a$ or $b = a$.
- Lastly, we turn to the notation for a limit ordinal λ. Suppose φ_e is a total computable function with values in $\mathcal{O}$ such that $\varphi_e(0) <_{\mathcal{O}} \varphi_e(1) <_{\mathcal{O}} \ldots$ and λ is the supremum of the sequence $(\lambda_n)_{n\in\mathbb{N}}$ given by $\lambda_n = |\varphi_e(n)|_{\mathcal{O}}$. We declare that $|3 \cdot 5^e|_{\mathcal{O}} = \lambda$ and declare that $b <_{\mathcal{O}} 3 \cdot 5^e$ if there is an n such that $b \leq_{\mathcal{O}} \varphi_e(n)$.

Note that we can carry out a transfinite induction on ordinal notation.

This set $\mathcal{O}$, generally referred to as *Kleene's* $\mathcal{O}$, has a number of interesting and useful properties. We list a few of them here:

- Each finite ordinal n has a unique notation $n_{\mathcal{O}}$, but each infinite ordinal has countably many notations.
- $\mathcal{O}$ is well founded because $<_{\mathcal{O}}$ induces a tree structure on $\mathcal{O}$.
- This tree structure enables us to think of a path through $\mathcal{O}$ along which each ordinal on the path has a unique notation.
- The ordinals with a notation in $\mathcal{O}$ are precisely the computable ordinals.

We now introduce a classification scheme for sets that extends the ideas inherent in the arithmetic hierarchy. We begin by recalling that the levels of the arithmetic hierarchy are defined by the complexity of the formulas required to define the sets at that level and that this complexity is calibrated in terms of the number and type of quantifier alternations in the set's defining formula. We note, though, that only number quantifiers were permitted. Now we turn our attention to second-order logic and permit set quantifiers as well.

At the lowest level of this new hierarchy, we have sets defined by formulas in second-order logic with no set quantifiers but any number of number quantifiers, bounded or not. These are categorized as $\Sigma^1_0 = \Pi^1_0 = \Delta^1_0$.

The next level of this hierarchy is all we will need for the purposes of this book, though the definition is inductive and can be extended. We say that a set is Σ^1_1 if it can be defined by a formula of the form

$$(\exists X_1)\ldots(\exists X_k)R(x; X_1, \ldots, X_k)$$

where R is a Σ_n relation for some n. If the set quantifiers are universal rather than existential, we say that the set is Π^1_1. The Δ^1_1 sets, or the sets in $\Sigma^1_1 \cap \Pi^1_1$, are called *hyperarithmetic*.

Now we can offer one final fact about $\mathcal{O}$: it is Π^1_1-complete.

References

1. Timothy H. McNicholl. Evaluative presentations. *J. Logic Comput.*, 35(5):Paper No. exaf036, 11, 2025.
2. Chris Eagle, Isaac Goldbring, Timothy McNicholl, and Russell Miller. Computable K-theory for C*-algebras: UHF algebras. *Trans. Amer. Math. Soc.*, to appear.
3. S. Barry Cooper. *Computability Theory*. Chapman & Hall/CRC, Boca Raton, FL, 2004.
4. Piergiorgio Odifreddi. *Classical Recursion Theory*. Number 125 in Studies in Logic and the Foundations of Mathematics. North-Holland, 1989.
5. Piergiorgio Odifreddi. *Classical Recursion Theory, Volume II*. Number 143 in Studies in Logic and the Foundations of Mathematics. North-Holland, 1999.
6. Rodney G. Downey and Denis R. Hirschfeldt. *Algorithmic Randomness and Complexity*. Springer, 2010.
7. C. J. Ash and J. Knight. *Computable Structures and the Hyperarithmetical Hierarchy*, volume 144 of *Studies in Logic and the Foundations of Mathematics*. North-Holland Publishing Co., Amsterdam, 2000.

Index

J. N. Y. Franklin et al., *Effective Metric Structure Theory*, Springer Monographs in Mathematics, https://doi.org/10.1007/978-3-032-13322-9

The manufacturer's authorised representative in the EU is Springer Nature Customer Service Centre GmbH, Europaplatz 3, 69115 Heidelberg, Germany. If you have any concerns regarding our products, please contact ProductSafety@springernature.com

Printed and bound by CPI Group (UK) Ltd, Croydon, CR0 4YY
07/10/2026
02229364-0002